Additional praise for *Fashionable Nonsense*

"Sokal is trying to stake out a territory free from the political claims of culture."
—Edward Rothstein, *The New York Times*

"The modern sciences are among the most remarkable of human achievements and cultural treasures. Like others, they merit—and reward—respectful and scrupulous engagement. Sokal and Bricmont show how easily such truisms can recede from view, and how harmful the consequences can be for intellectual life and human affairs. They also provide a thoughtful and constructive critical analysis of fundamental issues of empirical inquiry. It is a timely and substantial contribution."
—Noam Chomsky

"A brilliant and entertaining book...*Fashionable Nonsense* exposes the fraud."
—*The Advocate*

"A debut that promises to be [the debate's] most explosive incarnation yet."
—Kristina Zarlengo, *Salon Magazine*

"Sheer chutzpah and cleverness...The book is a sobering catalog of idiocies by some of those claimed to be the best thinkers of our times...I recommend this book."
—Russell Jacoby, *Los Angeles Weekly*

"[An] important and well-documented book...Every passage is followed by the authors' often humorous debunking of the writers' garbled science and obscure language. It's good reading."
—*Raleigh News-Observer*

"Their book has come like a breath of fresh air."
—John Weightman, *The Hudson Review*

"An in-depth examination."
—*Rolling Stone*

"Hilarious...What can be more irresistible than the opportunity to take some pompous, widely respected intellectual and knock him flat on his ass by exposing him as an idiot?"
—Fred Moody, *Seattle Weekly*

"[An] audacious debunking...The authors' fervor and the precision of their writing makes this a most engaging read."
—*Publishers Weekly*

"What they reveal is scandalous...true hilarity...The physicists aren't staging some sort of anti-theoretical pogrom; they're just standing up for rationality."
—Glenn Dixon, *Washington City Paper*

"This is a valuable and well-argued document in one of the key philosophical debates of our time."
—*Kirkus Reviews*

Also by Alan Sokal

Random Walks, Critical Phenomena,
and Triviality in Quantum Field Theory
(with Roberto Fernández and Jürg Frölich)

Fashionable Nonsense

Postmodern Intellectuals' Abuse of Science

Alan Sokal and Jean Bricmont

Picador
New York

www.picadorusa.com

Picador® is a U.S. registered trademark and is used by St. Martin's Press
under license from Pan Books Limited.

For information on Picador Reading Group Guides, as well as ordering,
please contact the Trade Marketing department at St. Martin's Press.
Phone: 1-800-221-7945 extension 763
Fax: 212-677-7456
E-mail: trademarketing@stmartins.com

Library of Congress Cataloging-in-Publication Data

Sokal, Alan.
    Fashionable nonsense : postmodern intellectuals' abuse of science
/ Alan Sokal and Jean Bricmont.
        p. cm.
    Includes bibliographical references and index
    ISBN 0-312-19445-1 (hc)
    ISBN 0-312-20407-8 (pbk)
    1. Science—Philosophy. I. Bricmont, J. (Jean) II. Title.

Q175.S3659 1998
501—dc21   98-35336
                                                                CIP

First published in France under the title *Impostures Intellectuelles* by
Editions Odile Jacob, 1997

10 9

For Marina

For Claire, Thomas, and Antoine

Contents

Preface to the English Edition

The publication in France of our book *Impostures Intellectuelles*[1] appears to have created a small storm in certain intellectual circles. According to Jon Henley in *The Guardian*, we have shown that "modern French philosophy is a load of old tosh."[2] According to Robert Maggiori in *Libération*, we are humorless scientistic pedants who correct grammatical errors in love letters.[3] We would like to explain briefly why neither is the case, and to answer both our critics and our over-enthusiastic supporters. In particular, we want to dispel a number of misunderstandings.

The book grew out of the now-famous hoax in which one of us published, in the American cultural-studies journal *Social Text*, a parody article crammed with nonsensical, but unfortunately authentic, quotations about physics and mathematics by prominent French and American intellectuals.[4] However, only a small fraction of the "dossier" discovered during Sokal's library research could be included in the parody. After showing this larger dossier to scientist and non-scientist friends, we became (slowly) convinced that it might be worth making it available to a wider audience. We wanted to explain, in non-technical terms, why the quotes are absurd or, in many cases, simply meaningless; and we wanted also to discuss the cultural circumstances

---

[1] Éditions Odile Jacob, Paris, October 1997.

[2] Henley (1997).

[3] Maggiori (1997).

[4] Sokal (1996a), reprinted here in Appendix A. The story of the hoax is described in more detail in Chapter 1 below.

that enabled these discourses to achieve such renown and to remain, thus far, unexposed.

But what exactly do we claim? Neither too much nor too little. We show that famous intellectuals such as Lacan, Kristeva, Irigaray, Baudrillard, and Deleuze have repeatedly abused scientific concepts and terminology: either using scientific ideas totally out of context, without giving the slightest justification—note that we are not against extrapolating concepts from one field to another, but only against extrapolations made without argument—or throwing around scientific jargon in front of their non-scientist readers without any regard for its relevance or even its meaning. We make no claim that this invalidates the rest of their work, on which we suspend judgment.

We are sometimes accused of being arrogant scientists, but our view of the hard sciences' role is in fact rather modest. Wouldn't it be nice (for us mathematicians and physicists, that is) if Gödel's theorem or relativity theory *did* have immediate and deep implications for the study of society? Or if the axiom of choice could be used to study poetry? Or if topology had something to do with the human psyche? But alas, it is not the case.

A second target of our book is epistemic relativism, namely the idea—which, at least when expressed explicitly, is much more widespread in the English-speaking world than in France—that modern science is nothing more than a "myth", a "narration" or a "social construction" among many others.[5] Besides some gross abuses (e.g. Irigaray), we dissect a number of confusions that are rather frequent in postmodernist and cultural-studies circles: for example, misappropriating ideas from the philosophy of science, such as the underdetermination of theory by evidence or the theory-ladenness of observation, in order to support radical relativism.

This book is therefore made up of two distinct—but related—works under one cover. First, there is the collection of

---

[5]Let us emphasize that our discussion is limited to epistemic/cognitive relativism; we do not address the more delicate issues of moral or aesthetic relativism.

extreme abuses discovered, rather haphazardly, by Sokal; this is the "fashionable nonsense" of our title. Second, there is our critique of epistemic relativism and of misconceptions about "postmodern science"; these analyses are considerably more subtle. The connection between these two critiques is primarily sociological: the French authors of the "nonsense" are fashionable in many of the same English-speaking academic circles where epistemic relativism is the coin of the realm.[6] There is also a weak logical link: if one accepts epistemic relativism, there is less reason to be upset by the misrepresentation of scientific ideas, which anyway are just another "discourse".

Obviously, we did not write this book just to point out some isolated abuses. We have larger targets in mind, but not necessarily those that are attributed to us. This book deals with mystification, deliberately obscure language, confused thinking, and the misuse of scientific concepts. The texts we quote may be the tip of an iceberg, but the iceberg should be defined as a set of intellectual practices, not a social group.

Suppose, for example, that a journalist discovers documents showing that several highly respected politicians are corrupt, and publishes them. (We emphasize that this is an analogy and that we do not consider the abuses described here to be of comparable gravity.) Some people will, no doubt, leap to the conclusion that *most* politicians are corrupt, and demagogues who stand to gain politically from this notion will encourage it.[7] But this extrapolation would be erroneous.

Similarly, to view this book as a generalized criticism of the humanities or the social sciences—as some French reviewers did—not only misunderstands our intentions, but is a curious assimilation, revealing a contemptuous attitude toward those

---

[6]This overlap is, however, not perfect. The French authors analyzed in this book are most fashionable, in the English-speaking world, in departments of literature, cultural studies and women's studies. Epistemic relativism is distributed rather more broadly, and is widespread also in domains of anthropology, education and sociology of science that exhibit little interest in Lacan or Deleuze.

[7]The politicians caught *in flagrante delicto* will also encourage this interpretation of the journalist's intentions, for different (but obvious) reasons.

fields in the minds of those reviewers.[8] As a matter of logic, either the humanities and social sciences are coterminous with the abuses denounced in this book, or they are not. If they are, then we would indeed be attacking those fields *en bloc*, but it would be justified. And if not (as we believe), there is simply no reason to criticize one scholar for what another in the same field says. More generally, any construal of our book as a blanket attack on X—whether X is French thought, the American cultural left or whatever—presupposes that the whole of X is permeated by the bad intellectual habits we are denouncing, and that charge has to be established by whoever makes it.

The debates sparked by Sokal's hoax have come to encompass an ever-wider range of ever-more-tenuously related issues, concerning not only the conceptual status of scientific knowledge or the merits of French poststructuralism, but also the social role of science and technology, multiculturalism and "political correctness", the academic left versus the academic right, and the cultural left versus the economic left. We want to emphasize that this book does *not* deal with most of these topics. In particular, the ideas analyzed here have little, if any, conceptual or logical connection with politics. Whatever one's views on Lacanian mathematics or the theory-ladenness of observation, one may hold, without fear of contradiction, any view whatsoever on military spending, national health insurance, or gay marriage. There is, to be sure, a *sociological* link—though its magnitude is often exaggerated—between the "postmodernist" intellectual currents we are criticizing and some sectors of the American academic left. Were it not for this link, we would not mention politics at all. But we do not want our book to be seen as one more shot in the dreary "Culture Wars", still less as one from the right. Critical thinking about the unfairness of our economic system and about racial and sexual op-

---

[8]Marc Richelle, in his very interesting and balanced book (1998), expresses the fear that some readers (and especially *non*-readers) of our book will jump to the conclusion that all the social sciences are nonsense. But he is careful to emphasize that this is not *our* view.

pression has grown in many academic institutions since the 1960s and has been subjected, in recent years, to much derision and unfair criticism. There is nothing in our book that can be construed, even remotely, in that genre.

Our book faces a quite different institutional context in France and in the English-speaking world. While the authors we criticize have had a profound impact on French higher education and have numerous disciples in the media, the publishing houses and the intelligentsia—hence some of the furious reactions to our book—their Anglo-American counterparts are still an embattled minority within intellectual circles (though a well-entrenched one in some strongholds). This tends to make them look more "radical" and "subversive" than they really are, both in their own eyes and in those of their critics. But our book is not against political radicalism, it is against intellectual confusion. Our aim is not to criticize the left, but to help defend it from a trendy segment of itself. Michael Albert, writing in *Z Magazine*, summarized this well: "There is nothing truthful, wise, humane, or strategic about confusing hostility to injustice and oppression, which is leftist, with hostility to science and rationality, which is nonsense."[9]

This edition is, in most respects, a straight translation from the French original. We have omitted a chapter on the misunderstandings of relativity by Henri Bergson and his successors, which seemed to us of marginal interest for most British and American readers.[10] Conversely, we have expanded a few discussions concerning intellectual debates in the English-speaking world. We have also made many small changes to improve the clarity of the original text, to correct minor imprecisions, and to forestall misunderstandings. We thank the many readers of the French edition who offered us their suggestions.

While writing this book, we have benefited from innumer-

---

[9]Albert (1996, p. 69). We shall return to these political issues in the Epilogue.

[10]Chapter 11 of the French original.

able discussions and debates and have received much encouragement and criticism. Although we are unable to thank individually all those who have contributed, we do want to express our gratitude to those who helped us by pointing out references or by reading and criticizing parts of the manuscript: Michael Albert, Robert Alford, Roger Balian, Louise Barre, Paul Boghossian, Raymond Boudon, Pierre Bourdieu, Jacques Bouveresse, Georges Bricmont, James Robert Brown, Tim Budden, Noam Chomsky, Helena Cronin, Bérangère Deprez, Jean Dhombres, Cryano de Dominicis, Pascal Engel, Barbara Epstein, Roberto Fernández, Vincent Fleury, Julie Franck, Allan Franklin, Paul Gérardin, Michel Gevers, Michel Ghins, Yves Gingras, Todd Gitlin, Gerald Goldin, Sylviane Goraj, Paul Gross, Étienne Guyon, Michael Harris, Géry-Henri Hers, Gerald Holton, John Huth, Markku Javanainen, Gérard Jorland, Jean-Michel Kantor, Noretta Koertge, Hubert Krivine, Jean-Paul Krivine, Antti Kupiainen, Louis Le Borgne, Gérard Lemaine, Geert Lernout, Jerrold Levinson, Norm Levitt, Jean-Claude Limpach, Andréa Loparic, John Madore, Christian Maes, Francis Martens, Tim Maudlin, Sy Mauskopf, Jean Mawhin, Maria McGavigan, N. David Mermin, Enrique Muñoz, Meera Nanda, Michael Nauenberg, Hans-Joachim Niemann, Marina Papa, Patrick Peccatte, Jean Pestieau, Daniel Pinkas, Louis Pinto, Patricia Radelet-de Grave, Marc Richelle, Benny Rigaux-Bricmont, Ruth Rosen, David Ruelle, Patrick Sand, Mónica Santoro, Abner Shimony, Lee Smolin, Philippe Spindel, Hector Sussmann, Jukka-Pekka Takala, Serge Tisseron, Jacques Treiner, Claire Van Cutsem, Jacques Van Rillaer, Loïc Wacquant, M. Norton Wise, Nicolas Witkowski, and Daniel Zwanziger. We are also indebted to our editors Nicky White and George Witte for many valuable suggestions. We emphasize that these people are not necessarily in agreement with the contents or even the intention of this book.

Finally, we thank Marina, Claire, Thomas, and Antoine for having put up with us for the past two years.

Fashionable Nonsense

1. Introduction

> So long as authority inspires awe, confusion and absurdity
> enhance conservative tendencies in society. Firstly, because clear
> and logical thinking leads to a cumulation of knowledge (of which
> the progress of the natural sciences provides the best example)
> and the advance of knowledge sooner or later undermines the
> traditional order. Confused thinking, on the other hand, leads
> nowhere in particular and can be indulged indefinitely without
> producing any impact upon the world.
> —*Stanislav Andreski, Social Sciences as Sorcery (1972, p. 90)*

The story of this book begins with a hoax. For some years, we have been surprised and distressed by the intellectual trends in certain precincts of American academia. Vast sectors of the humanities and the social sciences seem to have adopted a philosophy that we shall call, for want of a better term, "postmodernism": an intellectual current characterized by the more-or-less explicit rejection of the rationalist tradition of the Enlightenment, by theoretical discourses disconnected from any empirical test, and by a cognitive and cultural relativism that regards science as nothing more than a "narration", a "myth" or a social construction among many others.

To respond to this phenomenon, one of us (Sokal) decided to try an unorthodox (and admittedly uncontrolled) experiment: submit to a fashionable American cultural-studies journal, *Social Text*, a parody of the type of work that has proliferated in recent years, to see whether they would publish it. The article, entitled "Transgressing the Boundaries: Toward a Transformative Hermeneutics of Quantum Gravity"[1], is chock-full of absur-

---

[1] We reprint this article in Appendix A, followed by some brief comments in Appendix B.

dities and blatant non-sequiturs. In addition, it asserts an extreme form of cognitive relativism: after mocking the old-fashioned "dogma" that "there exists an external world, whose properties are independent of any individual human being and indeed of humanity as a whole", it proclaims categorically that "physical 'reality', no less than social 'reality', is at bottom a social and linguistic construct". By a series of stunning leaps of logic, it arrives at the conclusion that "the π of Euclid and the G of Newton, formerly thought to be constant and universal, are now perceived in their ineluctable historicity; and the putative observer becomes fatally de-centered, disconnected from any epistemic link to a space-time point that can no longer be defined by geometry alone". The rest is in the same vein.

And yet, the article was accepted and published. Worse, it was published in a special issue of *Social Text* devoted to rebutting the criticisms levelled against postmodernism and social constructivism by several distinguished scientists.[2] For the editors of *Social Text*, it was hard to imagine a more radical way of shooting themselves in the foot.

Sokal immediately revealed the hoax, provoking a firestorm of reaction in both the popular and academic press.[3] Many researchers in the humanities and social sciences wrote to Sokal, sometimes very movingly, to thank him for what he had done and to express their own rejection of the postmodernist and relativist tendencies dominating large parts of their disciplines. One student felt that the money he had earned to

---

[2]Among these criticisms, see for example Holton (1993), Gross and Levitt (1994), and Gross, Levitt, and Lewis (1996). The special issue of *Social Text* is introduced by Ross (1996). The parody is Sokal (1996a). The motivations for the parody are discussed in more detail in Sokal (1996c), which is reprinted here in Appendix C, and in Sokal (1997a). For earlier criticisms of postmodernism and social constructivism from a somewhat different political perspective—which are not, however, addressed in the *Social Text* issue—see e.g. Albert (1992–93), Chomsky (1992–93) and Ehrenreich (1992–93).

[3]The hoax was revealed in Sokal (1996b). The scandal landed (to our utter surprise) on the front page of the *New York Times* (Scott 1996), the *International Herald Tribune* (Landsberg 1996), the [London] *Observer* (Ferguson 1996), *Le Monde* (Weill 1996), and several other major newspapers. Among the reactions, see in particular the analyses by Frank (1996), Pollitt (1996), Willis (1996), Albert (1996), Weinberg (1996a, 1996b), Boghossian (1996), and Epstein (1997).

finance his studies had been spent on the clothes of an emperor who, as in the fable, was naked. Another wrote that he and his colleagues were thrilled by the parody, but asked that his sentiments be held in confidence because, although he wanted to help change his discipline, he could do so only after securing a permanent job.

But what was all the fuss about? Media hype notwithstanding, the mere fact the parody was published proves little in itself; at most it reveals something about the intellectual standards of *one* trendy journal. More interesting conclusions can be derived, however, by examining the *content* of the parody.[4] On close inspection, one sees that the parody was constructed around quotations from eminent French and American intellectuals about the alleged philosophical and social implications of mathematics and the natural sciences. The passages may be absurd or meaningless, but they are nonetheless authentic. In fact, Sokal's only contribution was to provide a "glue" (the "logic" of which is admittedly whimsical) to join these quotations together and praise them. The authors in question form a veritable pantheon of contemporary "French theory": Gilles Deleuze, Jacques Derrida, Félix Guattari, Luce Irigaray, Jacques Lacan, Bruno Latour, Jean-François Lyotard, Michel Serres, and Paul Virilio.[5] The citations also include many prominent American academics in Cultural Studies and related fields; but these authors are often, at least in part, disciples of or commentators on the French masters.

Since the quotations included in the parody were rather

---

[4]See Sokal (1998) for a more detailed discussion.

[5]In this book we have added Jean Baudrillard and Julia Kristeva to the list. Five of the ten "most important" French philosophers identified by Lamont (1987, note 4) are Baudrillard, Deleuze, Derrida, Lyotard, and Serres. Three of the six French philosophers chosen by Mortley (1991) are Derrida, Irigaray, and Serres. Five of the eight French philosophers interviewed by Rötzer (1994) are Baudrillard, Derrida, Lyotard, Serres, and Virilio. These same authors show up among the 39 Western thinkers interviewed by *Le Monde* (1984a, b), and one finds Baudrillard, Deleuze, Derrida, Irigaray, Kristeva, Lacan, Lyotard, and Serres among the 50 contemporary Western thinkers selected by Lechte (1994). Here the appellation "philosopher" is used in a broad sense; a more precise term would be "philosophico-literary intellectual".

brief, Sokal subsequently assembled a series of longer texts to illustrate these authors' handling of the natural sciences, which he circulated among his scientific colleagues. Their reaction was a mixture of hilarity and dismay: they could hardly believe that anyone—much less renowned intellectuals—could write such nonsense. However, when non-scientists read the material, they pointed out the need to explain, in lay terms, exactly *why* the cited passages are absurd or meaningless. From that moment, the two of us worked together to produce a series of analyses and commentaries on the texts, resulting in this book.

What We Intend to Show

The goal of this book is to make a limited but original contribution toward the critique of the admittedly nebulous Zeitgeist that we have called "postmodernism". We make no claim to analyze postmodernist thought in general; rather, our aim is to draw attention to a relatively little-known aspect, namely the repeated abuse of concepts and terminology coming from mathematics and physics. We shall also analyze certain confusions of thought that are frequent in postmodernist writings and that bear on either the content or the philosophy of the natural sciences.

The word "abuse" here denotes one or more of the following characteristics:

1) Holding forth at length on scientific theories about which one has, at best, an exceedingly hazy idea. The most common tactic is to use scientific (or pseudo-scientific) terminology without bothering much about what the words actually *mean*.

2) Importing concepts from the natural sciences into the humanities or social sciences without giving the slightest conceptual or empirical justification. If a biologist wanted to apply, in her research, elementary notions of mathematical topology, set theory or differential geometry, she would be asked to give some explanation. A vague analogy would not be taken very se-

riously by her colleagues. Here, by contrast, we learn from Lacan that the structure of the neurotic subject is exactly the torus (it is no less than reality itself, cf. p. 20), from Kristeva that poetic language can be theorized in terms of the cardinality of the continuum (p. 40), and from Baudrillard that modern war takes place in a non-Euclidean space (p. 147)—all without explanation.

3) Displaying a superficial erudition by shamelessly throwing around technical terms in a context where they are completely irrelevant. The goal is, no doubt, to impress and, above all, to intimidate the non-scientist reader. Even some academic and media commentators fall into the trap: Roland Barthes is impressed by the precision of Julia Kristeva's work (p. 38) and *Le Monde* admires the erudition of Paul Virilio (p. 169).

4) Manipulating phrases and sentences that are, in fact, meaningless. Some of these authors exhibit a veritable intoxication with words, combined with a superb indifference to their meaning.

These authors speak with a self-assurance that far outstrips their scientific competence: Lacan boasts of using "the most recent development in topology" (pp. 21–22) and Latour asks whether he has taught anything to Einstein (p. 131). They imagine, perhaps, that they can exploit the prestige of the natural sciences in order to give their own discourse a veneer of rigor. And they seem confident that no one will notice their misuse of scientific concepts. No one is going to cry out that the king is naked.

Our goal is precisely to say that the king is naked (and the queen too). But let us be clear. We are not attacking philosophy, the humanities or the social sciences *in general;* on the contrary, we feel that these fields are of the utmost importance and we want to warn those who work in them (especially students) against some manifest cases of charlatanism.[6] In particular, we want to "deconstruct" the reputation that certain texts have of

---

[6]If we refrain from giving examples of *good* work in these fields—as some readers have suggested—it is because making an exhaustive such list would go far beyond

being difficult because the ideas in them are so profound. In many cases we shall demonstrate that if the texts seem incomprehensible, it is for the excellent reason that they mean precisely nothing.

There are many different degrees of abuse. At one end, one finds extrapolations of scientific concepts, beyond their domain of validity, that are erroneous but for subtle reasons. At the other end, one finds numerous texts that are full of scientific words but entirely devoid of meaning. And there is, of course, a continuum of discourses that can be situated somewhere between these two extremes. Although we shall concentrate in this book on the most manifest abuses, we shall also briefly address some less obvious confusions concerning chaos theory (Chapter 7).

Let us stress that there is nothing shameful in being ignorant of calculus or quantum mechanics. What we are criticizing is the pretension of some celebrated intellectuals to offer profound thoughts on complicated subjects that they understand, at best, at the level of popularizations.[7]

At this point, the reader may naturally wonder: Do these abuses arise from conscious fraud, self-deception, or perhaps a combination of the two? We are unable to offer any categorical answer to this question, due to the lack of (publicly available) evidence. But, more importantly, we must confess that we do not find this question of great interest. Our aim here is to stimulate a critical attitude, not merely towards certain individuals, but towards a part of the intelligentsia (both in the United States and in Europe) that has tolerated and even encouraged this type of discourse.

---

our abilities, and a partial list would immediately bog us down in irrelevancies (why do you mention X and not Y?).

[7]Several commentators (Droit 1997, Stengers 1997, *Economist* 1997) have compared us to schoolteachers giving poor grades in mathematics and physics to Lacan, Kristeva *et al.* But the analogy is faulty: in school one is obliged to study certain subjects, but no one forced these authors to invoke technical mathematical concepts in their writings.

Yes, But . . .

Before proceeding any further, let us answer some of the objections that will no doubt occur to the reader:

1. *The quotations' marginality.* It could be argued that we are splitting hairs, criticizing authors who admittedly have no scientific training and who have perhaps made a mistake in venturing onto unfamiliar terrain, but whose contribution to philosophy and/or the social sciences is nevertheless important and is in no way invalidated by the "small errors" we have uncovered. We would respond, first of all, that these texts contain much more than mere "errors": they display a profound indifference, if not a disdain, for facts and logic. Our goal is not, therefore, to poke fun at literary critics who make mistakes when citing relativity or Gödel's theorem, but to defend the canons of rationality and intellectual honesty that are (or should be) common to all scholarly disciplines.

It goes without saying that we are not competent to judge the non-scientific aspects of these authors' work. We understand perfectly well that their "interventions" in the natural sciences do not constitute the central themes of their œuvre. But when intellectual dishonesty (or gross incompetence) is discovered in one part—even a marginal part—of someone's writings, it is natural to want to examine more critically the rest of his or her work. We do not want to prejudge the results of such an analysis, but simply to remove the aura of profundity that has sometimes intimidated students (and professors) from undertaking it.

When ideas are accepted on the basis of fashion or dogma, they are especially sensitive to the exposure even of marginal aspects. For example, geological discoveries in the eighteenth and nineteenth centuries showed that the earth is vastly older than the 5000-or-so years recounted in the Bible; and although these findings directly contradicted only a small part of the Bible, they had the indirect effect of undermining its overall credibility as a factual account of history, so that nowadays few people (except in the United States) believe in the Bible in the *literal* way that most Europeans did only a few centuries ago. Consider, by contrast, Isaac Newton's work: it is estimated that 90 percent of his writings deal with alchemy or mysticism. But, so what? The rest survives because it is based on solid empirical and rational arguments. Similarly, most of Descartes' physics is false, but some of the philosophical questions he raised are still pertinent today. If the same can be said for the work of our authors, then our findings have only marginal relevance. But if these writers have become international stars primarily for sociological rather than intellectual reasons, and in part because they are masters of language and can impress their audience with a clever abuse of sophisticated terminology—nonscientific as well as scientific—then the revelations contained in this essay may indeed have significant repercussions.

Let us emphasize that these authors differ enormously in their attitude toward science and the importance they give it. They should not be lumped together in a single category, and we want to warn the reader against the temptation to do so. For example, although the quotation from Derrida contained in Sokal's parody is rather amusing[8], it is a one-shot abuse; since there is no systematic misuse of (or indeed attention to) science in Derrida's work, there is no chapter on Derrida in this book. By contrast, the work of Serres is replete with more-or-less poetic allusions to science and its history; but his assertions, though extremely vague, are in general neither completely meaningless nor completely false, and so we have not discussed them here in detail.[9] Kristeva's early writings relied strongly (and abusively) on mathematics, but she abandoned this approach more than twenty years ago; we criticize them here because we consider them symptomatic of a certain intellectual style. The other authors, by contrast, have all invoked science extensively in

[8]The complete quote can be found in Derrida (1970, pp. 265–268).

[9]See, nevertheless, Chapter 11 and pp. 222, 262–63 for some examples of more manifest abuses in Serres' work.

their work. Latour's writings provide considerable grist for the mill of contemporary relativism and are based on an allegedly rigorous analysis of scientific practice. The works of Baudrillard, Deleuze, Guattari and Virilio are filled with seemingly erudite references to relativity, quantum mechanics, chaos theory, etc. So we are by no means splitting hairs in establishing that their scientific erudition is exceedingly superficial. Moreover, for several authors, we shall supply references to additional texts where the reader can find numerous further abuses.

2. *You don't understand the context.* Defenders of Lacan, Deleuze *et al.* might argue that their invocations of scientific concepts are valid and even profound, and that our criticisms miss the point because we fail to understand the context. After all, we readily admit that we do not always understand the rest of these authors' work. Mightn't we be arrogant and narrow-minded scientists, missing something subtle and deep?

We would respond, first of all, that when concepts from mathematics or physics are invoked in another domain of study, some argument ought to be given to justify their relevance. In all the cases cited here, we have checked that no such argument is provided, whether next to the excerpt we quote or elsewhere in the article or book.

Moreover, there are some "rules of thumb" that can be used to decide whether mathematics are being introduced with some real intellectual goal in mind, or merely to impress the reader. First of all, in cases of legitimate use, the author needs to have a good understanding of the mathematics he/she is purporting to apply—in particular, there should be no gross mistakes—and he/she should explain the requisite technical notions, as clearly as possible, in terms that will be understandable to the intended reader (who is presumably a non-scientist). Secondly, because mathematical concepts have precise meanings, mathematics is useful primarily when applied to fields in which the concepts likewise have more-or-less precise meanings. It is difficult to see how the mathematical notion of compact space can be applied fruitfully to something as ill-defined as the "space of *jouissance*" in psychoanalysis. Thirdly, one should be particularly suspicious when abstruse mathematical concepts (like the axiom of choice in set theory) that are used rarely, if at all, in physics—and certainly never in chemistry or biology—miraculously become relevant in the humanities or the social sciences.

3. *Poetic licence.* If a poet uses words like "black hole" or "degree of freedom" out of context and without really understanding their scientific meaning, it doesn't bother us. Likewise, if a science-fiction writer uses secret passageways in space-time in order to send her characters back to the era of the Crusades, it is purely a question of taste whether one likes or dislikes the technique.

By contrast, we insist that the examples cited in this book have nothing to do with poetic licence. These authors are holding forth, in utter seriousness, on philosophy, psychoanalysis, semiotics, or sociology. Their works are the subject of innumerable analyses, exegeses, seminars, and doctoral theses.[10] Their intention is clearly to produce theory, and it is on this ground that we criticize them. Moreover, their style is usually heavy and pompous, so it is highly unlikely that their goal is principally literary or poetic.

4. *The role of metaphors.* Some people will no doubt think that we are interpreting these authors too literally and that the passages we quote should be read as metaphors rather than as precise logical arguments. Indeed, in certain cases the "science" *is* undoubtedly intended metaphorically; but what is the purpose of these metaphors? After all, a metaphor is usually employed to clarify an unfamiliar concept by relating it to a more familiar one, not the reverse. Suppose, for example, that in a theoretical physics seminar we were to explain a very technical concept in quantum field theory by comparing it to the concept

---

[10]To illustrate more clearly that their claims are taken seriously in at least some parts of the English-speaking academy, we shall cite secondary works that analyze and elaborate, for example, Lacan's topology and mathematical logic, Irigaray's fluid mechanics, and Deleuze and Guattari's pseudo-scientific inventions.

of aporia in Derridean literary theory. Our audience of physicists would wonder, quite reasonably, what is the goal of such a metaphor—whether or not it is apposite—apart from displaying our own erudition. In the same way, we fail to see the advantage of invoking, even metaphorically, scientific concepts that one oneself understands only shakily when addressing a readership composed almost entirely of non-scientists. Might the goal be to pass off as profound a rather banal philosophical or sociological observation, by dressing it up in fancy scientific jargon?

5. *The role of analogies.* Many authors, including some of those discussed here, try to argue by analogy. We are by no means opposed to the effort to establish analogies between diverse domains of human thought; indeed, the observation of a valid analogy between two existing theories can often be very useful for the subsequent development of both. Here, however, we think that the analogies are between well-established theories (in the natural sciences) and theories too vague to be tested empirically (for example, Lacanian psychoanalysis). One cannot help but suspect that the function of these analogies is to hide the weaknesses of the vaguer theory.

Let us emphasize that a half-formulated theory—be it in physics, biology, or the social sciences—cannot be redeemed simply by wrapping it in symbols or formulae. The sociologist Stanislav Andreski has expressed this idea with his habitual irony:

> The recipe for authorship in this line of business is as simple as it is rewarding: just get hold of a textbook of mathematics, copy the less complicated parts, put in some references to the literature in one or two branches of the social studies without worrying unduly about whether the formulae which you wrote down have any bearing on the real human actions, and give your product a good-sounding title, which suggests that you have found a key to an exact science of collective behaviour. (Andreski 1972, pp. 129–130)

Andreski's critique was originally aimed at American quantitative sociology, but it is equally applicable to some of the texts
cited here, notably those of Lacan and Kristeva.

  6. *Who is competent?* We have frequently been asked the following question: You want to prevent philosophers from speaking about science because they don't have the requisite formal
training; but what qualifications do you have to speak of philosophy? This question betrays a number of misunderstandings.
First of all, we have no desire to prevent anyone from speaking
about anything. Secondly, the intellectual value of an intervention is determined by its content, not by the identity of the
speaker, much less by his or her diplomas.[11] Thirdly, there is an
asymmetry: we do not purport to judge Lacan's psychoanalysis,
Deleuze's philosophy, or Latour's concrete work in sociology.
We limit ourselves to their statements about the mathematical
and physical sciences or about elementary problems in the philosophy of science.

---

[11]The linguist Noam Chomsky illustrates this very well:

> In my own professional work I have touched on a variety of different fields. I've
> done work in mathematical linguistics, for example, without any professional
> credentials in mathematics; in this subject I am completely self-taught, and not
> very well taught. But I've often been invited by universities to speak on
> mathematical linguistics at mathematics seminars and colloquia. No one has
> ever asked me whether I have the appropriate credentials to speak on these
> subjects; the mathematicians couldn't care less. What they want to know is what
> I have to say. No one has ever objected to my right to speak, asking whether I
> have a doctor's degree in mathematics, or whether I have taken advanced
> courses in this subject. That would never have entered their minds. They want to
> know whether I am right or wrong, whether the subject is interesting or not,
> whether better approaches are possible—the discussion dealt with the subject,
> not with my right to discuss it.
>   But on the other hand, in discussion or debate concerning social issues or
> American foreign policy, Vietnam or the Middle East, for example, the issue is
> constantly raised, often with considerable venom. I've repeatedly been
> challenged on grounds of credentials, or asked, what special training do you
> have that entitles you to speak of these matters. The assumption is that people
> like me, who are outsiders from a professional viewpoint, are not entitled to
> speak on such things.
>   Compare mathematics and the political sciences—it's quite striking. In
> mathematics, in physics, people are concerned with what you say, not with your
> certification. But in order to speak about social reality, you must have the proper
> credentials, particularly if you depart from the accepted framework of thinking.
> Generally speaking, it seems fair to say that the richer the intellectual substance
> of a field, the less there is a concern for credentials, and the greater is the
> concern for content. (Chomsky 1979, pp. 6–7)

7. *Don't you too rely on argument from authority?* For if we assert that Lacan's mathematics are nonsense, how is the non-scientist reader to judge? Mustn't he or she take our word for it?

Not entirely. First of all, we have tried hard to provide detailed explanations of the scientific background, so that the non-specialist reader can appreciate *why* a particular assertion is erroneous or meaningless. We may not have succeeded in all cases: space is limited, and scientific pedagogy is difficult. The reader is perfectly entitled to reserve judgment in those cases where our explanation is inadequate. But, most importantly, it should be remembered that our criticism does *not* deal primarily with errors, but with the manifest *irrelevance* of the scientific terminology to the subject supposedly under investigation. In all the reviews, debates and private correspondence that have followed the publication of our book in France, no one has given even the slightest argument explaining how that relevance could be established.

8. *But these authors are not "postmodernist".* It is true that the French authors discussed in this book do not all regard themselves as "postmodernist" or "poststructuralist". Some of these texts were published prior to the emergence of these intellectual currents, and some of these authors reject any link with these currents. Moreover, the intellectual abuses criticized in this book are not homogeneous; they can be classified, very roughly, into two distinct categories, corresponding roughly to two distinct phases in French intellectual life. The first phase is that of extreme structuralism, extending through the early 1970s: the authors try desperately to give vague discourses in the human sciences a veneer of "scientificity" by invoking the trappings of mathematics. Lacan's work and the early writings of Kristeva fall into this category. The second phase is that of poststructuralism, beginning in the mid-1970s: here any pretense at "scientificity" is abandoned, and the underlying philosophy (to the extent one can be discerned) tends toward irrationalism or nihilism. The texts of Baudrillard, Deleuze and Guattari exemplify this attitude.

Furthermore, the very idea that there exists a distinctive category of thought called "postmodernist" is much less widespread in France than in the English-speaking world. If we nevertheless employ this term for convenience, it is because all the authors analyzed here are utilized as fundamental points of reference in English-language postmodernist discourse, and because some aspects of their writings (obscure jargon, implicit rejection of rational thought, abuse of science as metaphor) are common traits of Anglo-American postmodernism. In any case, the validity of our critiques can in no way depend on the use of a word; our arguments must be judged, for each author, independently of his or her link—be it conceptually justified or merely sociological—with the broader "postmodernist" current.

9. *Why do you criticize these authors and not others?* A long list of "others" has been suggested, both in print and in private correspondence: these include virtually all applications of mathematics to the social sciences (e.g. economics), physicists' speculations in popular books (e.g. Hawking, Penrose), sociobiology, cognitive science, information theory, the Copenhagen interpretation of quantum mechanics, and the use of scientific concepts and formulas by Hume, La Mettrie, D'Holbach, Helvetius, Condillac, Comte, Durkheim, Pareto, Engels, and sundry others.[12]

Let us begin by observing that this question is irrelevant to the validity or invalidity of our arguments; at best it can be used to cast aspersions on our intentions. Suppose there are other abuses as bad as those of Lacan or Deleuze; how would that justify the latter?

However, since the question of the grounds for our "selection" is so often asked, let us try to answer it briefly. First of all, we have no desire to write a ten-volume encyclopedia on "nonsense since Plato", nor do we have the competence to do so. Our scope is limited, firstly, to abuses in those scientific fields in

---

[12]See, for example, Lévy-Leblond (1997) and Fuller (1998).

which we can claim some expertise, namely mathematics and physics[13]; secondly, to abuses that are currently fashionable in influential intellectual circles; and thirdly, to abuses that have not previously been analyzed in detail. However, even within these constraints, we do not claim that our set of targets is exhaustive or that they constitute a "natural kind". Quite simply, Sokal stumbled on most of these texts in the course of writing his parody, and we decided, after reflection, that it was worth making them public.

Furthermore, we contend that there is a profound difference between the texts analyzed here and most of the other examples that have been suggested to us. The authors quoted in this book clearly do not have more than the vaguest understanding of the scientific concepts they invoke and, most importantly, they fail to give any argument justifying the relevance of these scientific concepts to the subjects allegedly under study. They are engaged in name-dropping, not just faulty reasoning. Thus, while it is very important to evaluate critically the uses of mathematics in the social sciences and the philosophical or speculative assertions made by natural scientists, these projects are different from—and considerably more subtle than—our own.[14]

A related question is:

10. *Why do you write a book on this and not on more serious issues? Is postmodernism such a great danger to civilization?* First of all, this is an odd question. Suppose someone discovers documents relevant to the history of Napoleon and writes a book about it. Would anyone ask him whether he thinks

---

[13]It would be interesting to attempt a similar project on the abuse of biology, computer science, or linguistics, but we leave that task to people more qualified than ourselves.

[14]Let us mention in passing two examples of the latter type of critique, authored by one of us: a detailed analysis of the popular books of Prigogine and Stengers dealing with chaos, irreversibility and the arrow of time (Bricmont 1995a), and a criticism of the Copenhagen interpretation of quantum mechanics (Bricmont 1995b). In our opinion Prigogine and Stengers give the educated public a distorted view of the topics they treat, but their abuses do not even come close to those analyzed in this book. And the deficiencies of the Copenhagen interpretation are vastly subtler.

this is a more important topic than World War II? His answer, and ours, would be that an author writes on a subject under two conditions: that he is competent and that he is able to contribute something original. His subject will not, unless he is particularly lucky, coincide with the most important problem in the world.

Of course we do not think that postmodernism is a great danger to civilization. Viewed on a global scale, it is a rather marginal phenomenon, and there are far more dangerous forms of irrationalism—religious fundamentalism, for instance. But we do think that the critique of postmodernism is worthwhile for intellectual, pedagogical, cultural and political reasons; we shall return to these themes in the Epilogue.

Finally, to avoid useless polemics and facile "refutations", let us emphasize that this book is not a right-wing pamphlet against left-wing intellectuals, or an American imperialist attack against the Parisian intelligentsia, or a simple know-nothing appeal to "common sense". In fact, the scientific rigor we are advocating often leads to results at odds with "common sense"; obscurantism, confused thinking, anti-scientific attitudes and the quasi-religious veneration of "great intellectuals" are in no way left-wing; and the attachment of part of the American intelligentsia to postmodernism demonstrates that the phenomenon is international. In particular, our critique is in no way motivated by the "theoretical nationalism and protectionism" that French writer Didier Eribon claims to detect in the work of some American critics.[15] Our aim is, quite simply, to denounce intellectual posturing and dishonesty, from wherever they come. If a significant part of the postmodernist "discourse" in contemporary American and British academia is of French origin, it is equally true that English-language intellectuals have long since given it an authentic home-grown flavor.[16]

---

[15]Eribon (1994, p. 70).

[16]We shall return to these cultural and political themes in the Epilogue.

Plan of This Book

The bulk of this book consists of an analysis of texts, author by author. For the convenience of non-specialist readers, we have provided, in footnotes, brief explanations of the relevant scientific concepts as well as references to good popular and semi-popular explanatory texts.

Some readers will no doubt think that we are taking these texts too seriously. That is true, in some sense. But since these texts *are* taken seriously by many people, we think that they deserve to be analyzed with the greatest rigor. In some cases we have quoted rather long passages, at the risk of boring the reader, in order to show that we have not misrepresented the meaning of the text by pulling sentences out of context.

In addition to abuses in the strict sense, we have also analyzed certain scientific and philosophical confusions that underlie much postmodernist thinking. First, we shall consider the problem of cognitive relativism, and show that a series of ideas coming from the history and philosophy of science do not have the radical implications that are often attributed to them (Chapter 4). Next we shall address several misunderstandings concerning chaos theory and so-called "postmodern science" (Chapter 7). Finally, in the Epilogue, we shall situate our critique in a wider cultural context.

Many of the texts quoted in this book originally appeared in French. Where a published English translation exists, we have most often used it (sometimes noting our corrections); it is cited in the bibliography, along with the original French source in brackets. In other cases, the translation is ours. We have endeavored to remain as faithful as possible to the original French, and in case of doubt we have reproduced the latter in brackets or even *in toto*. We assure the reader that if the passage seems incomprehensible in English, it is because the original French is likewise.

2. Jacques Lacan

Lacan finally gives Freud's thought the scientific concepts it requires.
—*Louis Althusser, Écrits sur la psychanalyse (1993, p. 50)*

Lacan is, as he himself says, a crystal-clear author.
—*Jean-Claude Milner, L'œuvre claire (1995, p. 7)*

Jacques Lacan was one of the most famous and influential psychoanalysts of this century. Each year, dozens of books and articles are devoted to the analysis of his work. According to his disciples, he revolutionized the theory and practice of psychoanalysis; according to his critics, he is a charlatan and his writings are pure verbiage. We shall not enter here into the debate concerning the purely psychoanalytic part of Lacan's work. Rather, we shall limit ourselves to an analysis of his frequent references to mathematics, and show that Lacan illustrates perfectly, in different parts of his œuvre, the abuses listed in our introduction.

"Psychoanalytic Topology"

Lacan's mathematical interests centered primarily on topology, the branch of mathematics dealing (among other things) with the properties of geometrical objects—surfaces, solids, and so forth—that remain unchanged when the object is deformed without being torn. (According to the classic joke, a topologist is unable to tell a doughnut from a coffee cup, as both are solid objects with a single hole.) Lacan's writings contained some references to topology already in the 1950s; but the first extended (and publicly available) discussion goes back to a celebrated conference on *The Languages of Criticism and the Sciences of Man*, held at Johns Hopkins University in 1966. Here is an excerpt from Lacan's lecture:

> This diagram [the Möbius strip[17]] can be considered the basis of a sort of essential inscription at the origin, in the knot which constitutes the subject. This goes much further than you may think at first, because you can search for the sort of surface able to receive such inscriptions. You can perhaps see that the sphere, that old symbol for totality, is unsuitable. A torus, a Klein bottle, a cross-cut surface[18], are able to receive such a cut. And this diversity is very important as it explains many things about the structure of mental disease. If one can symbolize the subject by this fundamental cut, in the same way one can show that a cut on a torus corresponds to the neurotic subject, and on a cross-cut surface to another sort of mental disease. (Lacan 1970, pp. 192–193)

Perhaps the reader is wondering what these different topological objects have to do with the structure of mental disease. Well, so are we; and the rest of Lacan's text does nothing to clarify the matter. Nevertheless, Lacan insists that his topology "explains many things". In the discussion following his lecture, one finds the following dialogue:

> HARRY WOOLF: May I ask if this fundamental arithmetic and this topology are not in themselves a myth or merely at best an analogy for an explanation of the life of the mind?

---

[17]A Möbius strip can be constructed taking a rectangular strip of paper, twisting one of the short sides by 180 degrees, and gluing it to the other short side. In this way, one produces a surface with only one face: "front" and "back" are connected by a continuous path.

[18]A torus is the surface formed by a hollow tire. A Klein bottle is rather like a Möbius strip, but without an edge; to represent it concretely, one needs a Euclidean space of dimension at least four. The cross-cap (here called "cross-cut", probably due to a transcription error) is yet another type of surface.

JACQUES LACAN: Analogy to what? "S" designates something which can be written exactly as this S. And I have said that the "S" which designates the subject is instrument, matter, to symbolize a loss. A loss that you experience as a subject (and myself also). In other words, this gap between one thing which has marked meanings and this other thing which is my actual discourse that I try to put in the place where you are, you as not another subject but as people that are able to understand me. Where is the analogon? Either this loss exists or it doesn't exist. If it exists it is only possible to designate the loss by a system of symbols. In any case, the loss does not exist before this symbolization indicates its place. It is not an analogy. It is really in some part of the realities, this sort of torus. This torus really exists and it is exactly the structure of the neurotic. It is not an analogon; it is not even an abstraction, because an abstraction is some sort of diminution of reality, and I think it is reality itself. (Lacan 1970, pp. 195–196)

Here again, Lacan gives no argument to support his peremptory assertion that the torus "is exactly the structure of the neurotic" (whatever this means). Moreover, when asked explicitly whether it is simply an analogy, he denies it.

As the years passed, Lacan became increasingly fond of topology. A text from 1972 begins by playing on the etymology of the word (Greek *topos*, place + *logos*, word):

In this space of jouissance, to take something that is bounded, closed [*borné, fermé*] constitutes a locus [*lieu*], and to speak of it constitutes a topology. (Lacan 1975a, p. 14; Lacan 1998, p. 9; seminar originally held in 1972[19])

In this sentence, Lacan has used four technical terms from mathematical analysis (*space, bounded, closed, topology*) but

---

[19]We have here corrected the translation of the word *borné*, which in the mathematical context means "bounded".

without paying attention to their *meaning;* the sentence is meaningless from a mathematical point of view. Furthermore— and most importantly—Lacan never explains the relevance of these mathematical concepts for psychoanalysis. Even if the concept of *"jouissance"* had a clear and precise meaning, Lacan provides no reason whatsoever to think that *jouissance* can be considered a "space" in the technical sense of this word in topology. Nevertheless, he continues:

> In a text soon to be published that is at the cutting edge of my discourse last year, I believe I demonstrate the strict equivalence between topology and structure.[20] If we take that as our

---

[20]According to the translator's footnote as well as Roustang (1990, p. 87), the reference to "my discourse [from] last year" is to Lacan (1973). We have therefore reread this article and searched for the promised "demonstration" of "the strict equivalence between topology and structure". Now, the article contains long (and frankly bizarre) meditations mixing topology, logic, psychoanalysis, Greek philosophy, and virtually everything else under the kitchen sink—we shall quote a brief excerpt below, see pp. 32–36—but concerning the alleged equivalence between topology and "structure", one finds only the following:

> Topology is not "made to guide us" in structure. This structure is it—as retroaction of the chain order of which language consists.
> Structure is the aspherical concealed in the articulation of language insofar as an effect of subject takes hold of it.
> It is clear that, as far as meaning is concerned, this "takes hold of it" of the sub-sentence—pseudo-modal—reverberates from the object itself which it wraps, as verb, in its grammatical subject, and that there is a false effect of meaning, a resonance of the imaginary induced by the topology, according to whether the effect of subject makes a whirlwind of asphere [*sic*] or the subjective of this effect "reflects" itself from it.
> Here one must distinguish the ambiguity that inscribes itself from the meaning, that is, from the loop of the cut, and the suggestion of hole, that is, of structure, which makes sense of this ambiguity. (Lacan 1973, p. 40)

[Because Lacan's language is so obscure, we reproduce also the original French text:]

> La topologie n'est pas "faite pour nous guider" dans la structure. Cette structure, elle l'est—comme rétroaction de l'ordre de chaîne dont consiste le langage.
> La structure, c'est l'asphérique recelé dans l'articulation langagière en tant qu'un effet de sujet s'en saisit.
> Il est clair que, quant à la signification, ce "s'en saisit" de la sous-phrase, pseudo-modale, se répercute de l'objet même que comme verbe il enveloppe dans son sujet grammatical, et qu'il y a faux effet de sens, résonance de l'imaginaire induit de la topologie, selon que l'effet de sujet fait tourbillon d'asphère ou que le subjectif de cet effet s'en "réfléchit".

guide, what distinguishes anonymity from what we talk about
as jouissance—namely, what is regulated by law—is a geometry. A geometry implies the heterogeneity of locus, namely
that there is a locus of the Other.[21] Regarding this locus of the
Other, of one sex as Other, as absolute Other, what does the
most recent development in topology allow us to posit?

I will posit here the term "compactness."[22] Nothing is more
compact than a fault [*faille*], assuming that the intersection of
everything that is enclosed therein is accepted as existing over
an infinite number of sets, the result being that the intersection
implies this infinite number. That is the very definition of compactness. (Lacan 1975a, p. 14; Lacan 1998, p. 9)

Not at all: although Lacan uses quite a few key words from the
mathematical theory of compactness (see note 22), he mixes

Il y a ici à distinguer l'ambiguïté qui s'inscrit de la signification, soit de la boucle
de la coupure, et la suggestion de trou, c'est-à-dire de structure, qui de cette
ambiguïté fait sens. (Lacan 1973, p. 40)

If we leave aside Lacan's mystifications, the relationship between topology and
structure is easy to understand, but it depends upon what one means by "structure".
If this term is understood broadly—that is, as including linguistic and social
structures as well as mathematical structures—then it clearly cannot be reduced to
the purely mathematical notion of "topology". If, on the other hand, one understands
"structure" in its strictly mathematical sense, then one sees easily that topology is *one*
type of structure, but that there exist many others: order structure, group structure,
vector-space structure, manifold structure, etc.

[21]If the last two sentences have a meaning, they have, in any case, nothing to do with
geometry.

[22]Compactness is an important technical concept in topology, but rather difficult to
explain. Suffice it to say that in the nineteenth century, mathematicians (Cauchy,
Weierstrass, and others) put mathematical analysis on a solid basis by giving a precise
meaning to the concept of *limit*. These limits were initially used for sequences of *real*
numbers, but it was slowly realized that the notion of limit should be extended to
spaces of functions (for example, to study differential or integral equations).
Topology was born circa 1900 in part through these studies. Now, among *topological*
spaces one may distinguish a subclass called *compact spaces*, namely those in which
every *sequence* of elements possesses a *subsequence* having a limit. (Here we have
simplified somewhat, by limiting ourselves to *metric spaces*.) Another definition
(which can be proven to be equivalent to the first one) relies on the *intersection*
properties of *infinite* collections of *closed* sets. In the special case of *subsets* of
finite-dimensional Euclidean spaces, a set is compact if and only if it is *closed* and
bounded. Let us emphasize that all the italicized words above are technical terms
having very precise definitions (which in general are based on a long chain of other
definitions and theorems).

them up arbitrarily and without the slightest regard for their meaning. His "definition" of compactness is not just false: it is gibberish. Moreover, this "most recent development in topology" goes back to 1900–1930.

He continues as follows:

> The intersection I am talking about is the same one I put forward earlier as being that which covers or poses an obstacle to the supposed sexual relationship.
>
> Only "supposed," since I state that analytic discourse is premised solely on the statement that there is no such thing, that it is impossible to found [*poser*] a sexual relationship. Therein lies analytic discourse's step forward and it is thereby that it determines the real status of all the other discourses.
>
> Named here is the point that covers the impossibility of the sexual relationship as such. Jouissance, qua sexual, is phallic—in other words, it is not related to the Other as such.
>
> Let us follow here the complement of the hypothesis of compactness.
>
> A formulation is given to us by the topology I qualified as the most recent that takes as its point of departure a logic constructed on the investigation of numbers and that leads to the institution of a locus, which is not that of a homogeneous space. Let us take the same bounded[23], closed, supposedly instituted space—the equivalent of what I earlier posited as an intersection extending to infinity. If we assume it to be covered with open sets, in other words, sets that exclude their own limits—the limit is that which is defined as greater than one point and less than another, but in no case equal either to the point of departure or the point of arrival, to sketch it for you quickly[24]—it can be shown that it is equivalent to say that

---

[23]See note 19 above.

[24]In this sentence, Lacan gives an incorrect definition of *open set* and a meaningless "definition" of *limit*. But these are minor points compared to the overall confusion of the discourse.

the set of these open spaces always allows of a subcovering of open spaces, constituting a finity [*finitude*], namely, that the series of elements constitutes a finite series.

You may note that I did not say that they are countable. And yet that is what the term "finite" implies. In the end, we count them one by one. But before we can count them, we must find an order in them and we cannot immediately assume that that order is findable.[25]

What is implied, in any case, by the demonstrable finity of the open spaces that can cover the space that is bounded[26] and closed in the case of sexual jouissance? What is implied is that the said spaces can be taken one by one [*un par un*]— and since I am talking about the other pole, let us put this in the feminine—*une par une.*

That is the case in the space of sexual jouissance, which thereby proves to be compact. (Lacan 1975a, pp. 14–15; Lacan 1998, pp. 9–10)

This passage illustrates perfectly two "faults" in Lacan's discourse. Everything is based—at best—on analogies between topology and psychoanalysis that are unsupported by any argument. But, in fact, even the mathematical statements are devoid of meaning.

In the mid-1970s, Lacan's topological preoccupations shifted towards knot theory: see, for example, Lacan (1975a, pp. 107–123; 1998, pp. 122–136) and especially Lacan (1975b–e). For a detailed history of his obsessions with topology, see Roudinesco (1997, chapter 28). Lacan's disciples have given full accounts of his *topologie psychanalytique:* see, for example, Granon-Lafont (1985, 1990), Vappereau (1985, 1995), Nasio (1987, 1992), Darmon (1990) and Leupin (1991).

---

[25]This paragraph is pure pedantry. Obviously, if a set is finite, one can, in principle, "count" it and "order" it. All the discussions in mathematics concerning countability (see note 38 below) or the possibility of ordering sets are motivated by *infinite* sets.

[26]See note 19 above.

Imaginary Numbers

Lacan's predilection for mathematics is by no means marginal in his work. Already in the 1950s, his writings were full of graphs, formulas and "algorithms". Let us quote, by way of illustration, this excerpt from a seminar held in 1959:

> If you'll permit me to use one of those formulas which come to me as I write my notes, human life could be defined as a calculus in which zero was irrational. This formula is just an image, a mathematical metaphor. When I say "irrational," I'm referring not to some unfathomable emotional state but precisely to what is called an imaginary number. The square root of minus one doesn't correspond to anything that is subject to our intuition, anything real—in the mathematical sense of the term—and yet, it must be conserved, along with its full function. (Lacan 1977a, pp. 28–29, seminar held originally in 1959)

In this quote, Lacan confuses irrational numbers with imaginary numbers, while claiming to be "precise". They have nothing to do with one another.[27] Let us emphasize that the mathematical meanings of the words "irrational" and "imaginary" are quite distinct from their ordinary or philosophical meanings. To be sure, Lacan speaks here prudently of a metaphor, though it is hard to see what theoretical role this metaphor (human life as a "calculus in which zero was irrational") could fulfill. Nevertheless, a year later, he further developed the psychoanalytic role of imaginary numbers:

> Personally, I will begin with what is articulated in the sigla S(Ø) by being first of all a signifier. . . .

---

[27]A number is called *irrational* if it cannot be written as a ratio of two integers: for example, the square root of two, or π. (By contrast, zero is an integer, hence unavoidably a *rational* number.) The *imaginary* numbers, on the other hand, are introduced as solutions of polynomial equations that have no solutions among the real numbers: for example, $x^2 + 1 = 0$, one of whose solutions is denoted $i = \sqrt{-1}$ and the other $-i$.

And since the battery of signifiers, as such, is by that very fact complete, this signifier can only be a line [*trait*] that is drawn from its circle without being able to be counted part of it. It can be symbolized by the inherence of a (−1) in the whole set of signifiers.

As such it is inexpressible, but its operation is not inexpressible, for it is that which is produced whenever a proper noun is spoken. Its statement equals its signification.

Thus, by calculating that signification according to the algebraic method used here, namely:

$$\frac{\text{S (signifier)}}{\text{s (signified)}} = \text{s (the statement), with S} = (-1), \text{ produces:}$$
$$s = \sqrt{-1}.$$

(Lacan 1977b, pp. 316–317, seminar originally held in 1960)

Here Lacan can only be pulling the reader's leg. Even if his "algebra" had a meaning, the "signifier", "signified" and "statement" that appear within it are obviously not numbers, and his horizontal bar (an arbitrarily chosen symbol) does not denote the division of two numbers. Therefore, his "calculations" are pure fantasies.[28] Nevertheless, two pages later, Lacan returns to the same theme:

No doubt Claude Lévi-Strauss, in his commentary on Mauss, wished to recognize in it the effect of a zero symbol. But it seems to me that what we are dealing with here is rather the signifier of the lack of this zero symbol. That is why, at the risk of incurring a certain amount of opprobrium, I have indicated to what point I have pushed the distortion of the mathematical algorithm in my use of it: the symbol $\sqrt{-1}$, which is still written as '*i*' in the theory of complex numbers, is obviously justified only because it makes no claim to any automatism in its later use.

---

[28]For an exegesis of Lacan's "algorithm" that is almost as ridiculous as the original text, see Nancy and Lacoue-Labarthe (1992, part I, chapter 2).

. . .

Thus the erectile organ comes to symbolize the place of *jouissance*, not in itself, or even in the form of an image, but as a part lacking in the desired image: that is why it is equivalent to the $\sqrt{-1}$ of the signification produced above, of the *jouissance* that it restores by the coefficient of its statement to the function of lack of signifier (-1). (Lacan 1977b, pp. 318–320)

It is, we confess, distressing to see our erectile organ equated to $\sqrt{-1}$. This reminds us of Woody Allen, who, in *Sleeper*, objects to the reprogramming of his brain: "You can't touch my brain, it's my second-favorite organ!"

Mathematical Logic

In some of his texts, Lacan does less violence to mathematics. For example, in the quote below, he mentions two fundamental problems in the philosophy of mathematics: the nature of mathematical objects, in particular of the natural numbers (1, 2, 3, . . .), and the validity of reasoning by "mathematical induction" (if a property is true for the number 1 and if one can show that its truth for the number n implies its truth for the number $n + 1$, then one can deduce that the property is true for all natural numbers).

After fifteen years I have taught my pupils to count at most up to five which is difficult (four is easier) and they have understood that much. But for tonight permit me to stay at two. Of course, what we are dealing with here is the question of the integer, and the question of integers is not a simple one as I think many people here know. It is only necessary to have, for instance, a certain number of sets and a one-to-one correspondence. It is true for example that there are exactly as many people sitting in this room as there are seats. But it is necessary to have a collection composed of integers to constitute an integer, or what is called a natural number. It is, of course, in part natural but only in the sense that we do not understand why it exists. Counting is not an empirical fact and it is impossible to deduce the act of counting from empirical data alone. Hume tried but Frege demonstrated perfectly the ineptitude of the attempt. The real difficulty lies in the fact that every integer is in itself a unit. If I take two as a unit, things are very enjoyable, men and women for instance—love plus unity! But after a while it is finished, after these two there is nobody, perhaps a child, but that is another level and to generate three is another affair. When you try to read the theories of mathematicians regarding numbers you find the formula "*n* plus 1" ($n + 1$) as the basis of all the theories. (Lacan 1970, pp. 190–191)

So far, this is not too bad: those who already know the subject can recognize the vague allusions to classic debates (Hume/Frege, mathematical induction) and separate them from some rather questionable statements (for example, what does it mean to say "The real difficulty lies in the fact that every integer is in itself a unit"?). But from here on, Lacan's reasoning becomes increasingly obscure:

It is this question of the "one more" that is the key to the genesis of numbers and instead of this unifying unity that constitutes two in the first case I propose that you consider the real numerical genesis of two.

It is necessary that this two constitute the first integer which is not yet born as a number before the two appears. You have made this possible because the *two* is here to grant existence to the first *one:* put *two* in the place of *one* and consequently in the place of the *two* you see *three* appear. What we have here is something which I can call the *mark*. You already have something which is marked or something which is not marked. It is with the first mark that we have the status of the thing. It is exactly in this fashion that Frege explains the genesis of the number; the class which is characterized by no elements is the first class; you have one at the place of zero and afterward it is easy to understand how the place of one becomes the second place which makes place for two, three, and so on.[29] (Lacan 1970, p. 191, italics in the original)

And it is at this moment of obscurity that Lacan introduces, without explanation, the alleged link with psychoanalysis:

The question of the two is for us the question of the subject, and here we reach a fact of psychoanalytical experience in as much as the two does not complete the one to make two, but must repeat the one to permit the one to exist. This first repetition is the only one necessary to explain the genesis of the number, and only one repetition is necessary to constitute the status of the subject. The unconscious subject is something that tends to repeat itself, but only one such repetition is necessary to constitute it. However, let us look more precisely at what is necessary for the second to repeat the first in order that we may have a repetition. This question cannot be answered too quickly. If you answer too quickly, you will answer that it is necessary that they are the same. In this case the principle of the two would be that of twins—and why not triplets or quintuplets? In my day we used to teach children that they must not add, for instance, microphones with dictionaries; but this is absolutely absurd, because we would not have addition if we were not able to add microphones with dictionaries or as Lewis Carroll says, cabbages with kings.

---

[29] This last sentence may be a rather confused allusion to a technical procedure used in mathematical logic to define the natural numbers in terms of sets: 0 is identified with the empty set $\emptyset$ (i.e. the set having no element); then 1 is identified with the set $\{\emptyset\}$ (i.e. the set having $\emptyset$ as its sole element); then 2 is identified with the set $\{\emptyset, \{\emptyset\}\}$ (i.e. the set having the two elements $\emptyset$ and $\{\emptyset\}$); and so forth.

The sameness is not in *things* but in the *mark* which makes it possible to add things with no consideration as to their differences. The mark has the effect of rubbing out the difference, and this is the key to what happens to the subject, the unconscious subject in the repetition; because you know that this subject repeats something peculiarly significant, the subject is here, for instance, in this obscure thing that we call in some cases trauma, or exquisite pleasure. (Lacan 1970, pp. 191–192, italics in the original)

Thereafter, Lacan tries to link mathematical logic and linguistics:

I have only considered the beginning of the series of the integers, because it is an intermediary point between language and reality. Language is constituted by the same sort of unitary traits that I have used to explain the one and the one more. But this trait in language is not identical with the unitary trait, since in language we have a collection of differential traits. In other words, we can say that language is constituted by a set of signifiers—for example, *ba*, *ta*, *pa*, etc., etc.—a set which is finite. Each signifier is able to support the same process with regard to the subject, and it is very probable that the process of the integers is only a special case of this relation between signifiers. The definition of this collection of signifiers is that they constitute what I call the Other. The difference afforded by the existence of language is that each signifier (contrary to the unitary trait of the integer number) is, in most cases, not identical with itself—precisely because we have a collection of signifiers, and in this collection one signifier may or may not designate itself. This is well known and is the principle of Russell's paradox. If you take the set of all elements which are not members of themselves,

$$x \notin x$$

the set that you constitute with such elements leads you to a paradox which, as you know, leads to a contradiction.[30] In simple terms, this only means that in a universe of discourse nothing contains everything[31], and here you find again the gap that constitutes the subject. The subject is the introduction of a loss in reality, yet nothing can introduce that, since by status reality is as full as possible. The notion of a loss is the effect afforded by the instance of the trait which is what, with the intervention of the letter you determine, places—say a_1 a_2 a_3—and the places are spaces, for a lack. (Lacan 1970, p. 193)

Firstly, from the moment that Lacan claims to speak "in simple terms", everything becomes obscure. Secondly—and most importantly—no argument is given to link these paradoxes belonging to the foundations of mathematics with "the gap that constitutes the subject" in psychoanalysis. Might Lacan be trying to impress his audience with a superficial erudition?

Overall, this text illustrates perfectly the second and third abuses on our list: Lacan shows off, to non-experts, his knowledge in mathematical logic; but his account is neither original nor pedagogical from a mathematical point of view, and the link with psychoanalysis is not supported by any argument.[32]

In other texts, even the supposedly "mathematical" content is meaningless. For example, in an article written in 1972, Lacan

---

[30] The paradox to which Lacan is alluding here is due to Bertrand Russell (1872–1970). Let us begin by observing that most "normal" sets do not contain themselves as an element: for example, the set of all chairs is not itself a chair, the set of all whole numbers is not a whole number, etc. On the other hand, some sets do apparently contain themselves as an element: for example, the set of all abstract ideas is itself an abstract idea, the set of all sets is a set, etc. Consider now the set of all sets that do not contain themselves as an element. Does it contain itself? If the answer is yes, then it cannot belong to the set of all sets that *do not contain* themselves, and therefore the answer should be no. But if the answer is no, then it *must* belong to the set of all sets that do not contain themselves, and the answer should be yes. To escape from this paradox, logicians have replaced the naive concept of set by a variety of axiomatic theories.

[31] This is perhaps an allusion to a *different* (though related) paradox, due to Georg Cantor (1845–1918), concerning the nonexistence of the "set of all sets".

[32] See e.g. Miller (1977/78) and Ragland-Sullivan (1990) for worshipful commentary on Lacan's mathematical logic.

states his famous maxim—"there is no sexual relation"—and translates this obvious truth in his famous "formulae of sexuation"[33]:

> Everything can be held to develop itself around what I set forth about the logical correlation of two formulas that, to be inscribed mathematically $\forall x \cdot \Phi x$, and $\exists x \cdot \overline{\Phi x}$, can be stated as[34]:
>
> the first, for all x, Φx is satisfied, which can be translated by a T denoting truth value. This, translated into the analytic discourse of which it is the practice to make sense, "means" that every subject as such—that being what is at stake in this discourse—inscribes itself in the phallic function in order to ward off the absence of the sexual relation (the practice of making sense is exactly to refer to this ab-sense);
>
> the second, there is by exception the case, familiar in mathematics (the argument $x = 0$ in the exponential function $1/x$), the case where there exists an x for which Φx, the function, is not satisfied, i.e. does not function, is in fact excluded.
>
> It is precisely from there that I conjugate the all of the universal, more modified than one imagines in the *forall* of the quantor, to the *there exists one* with which the quantic pairs it off, its difference being patent with what is implied by the proposition that Aristotle calls particular. I conjugate them of what the *there exists one* in question, to make a limit on the *forall*, is what affirms or confirms it (what a proverb already objects to Aristotle's contradictory).
>
> . . .
>
> That I state the existence of a subject to posit it of a saying no to the propositional function Φx, implies that it inscribes itself of a quantor of which this function finds itself cut

---

[33]Because Lacan's language is so obscure and frequently ungrammatical, we have reproduced the complete French text following our best attempt at a translation.

[34]In mathematical logic, the symbol $\forall x$ means "for all x", and the symbol $\exists x$ means "there exists at least one x such that"; they are called the "universal quantifier" and the "existential quantifier", respectively. Further down in the text, Lacan writes Ax and Ex to denote the same concepts.

off from the fact that it has at this point no value that one can denote truth value, which means no error either, the false only to understand *falsus* as fallen, which I already emphasized.

In classical logic, to think of it, the false is not seen only as being of truth the reverse, it designates truth as well.

It is thus correct to write as I do: $\mathrm{E}x \cdot \overline{\Phi x}$.

. . .

That the subject here proposes itself to be called woman depends on two modes. Here they are:

$$\overline{\mathrm{E}x} \cdot \overline{\Phi x}; \text{ and } \overline{\mathrm{A}x} \cdot \Phi x.$$

Their inscription is not used in mathematics.[35] To deny, as the bar put above the quantor indicates, to deny that *there exists one* is not done, much less that the *forall* should notforall itself.

It is there, however, that the meaning of the saying delivers itself, of that which, conjugating the *nyania* that noises the sexes in company, it makes up for the fact that, between them, the relation isn't.

Which is to be understood not in the sense that, to reduce our quantors to their reading according to Aristotle, would set the *notexistone* equal to the *noneis* of its negative universal, would make the μή πάντες come back, the *notall* (that he was nevertheless able to formulate), to testify to the existence of a subject to say no to the phallic function, that to suppose it of the contrariety said of two particulars.

This is not the meaning of the saying, which inscribes itself of these quantors.

It is: that in order to introduce itself as a half to say about women, the subject determines itself from the fact that, since

---

[35]Just so. The bar ‾‾‾‾ denotes negation ("it is false that") and can thus be applied only to complete propositions, not to isolated quantifiers such as $\mathrm{E}x$ or $\mathrm{A}x$. One might think that here Lacan means $\overline{\mathrm{E}x} \cdot \overline{\Phi x}$ and $\overline{\mathrm{A}x} \cdot \Phi x$—which would in fact be logically equivalent to his starting propositions $\mathrm{A}x \cdot \Phi x$ and $\mathrm{E}x \cdot \overline{\Phi x}$—but he makes clear that this banal rewriting is *not* his intention. Anyone is free to introduce a new notation, but he then has the obligation of explaining its meaning.

there does not exist a suspension of the phallic function, everything can here be said of it, even if it comes from the without-reason. But it is an out-of-universe whole, which is read without a hitch from the second quantor as *notall*.

The subject in the half where it determines itself from the denied quantors, it is that nothing existing could put a limit on the function, that could not assure itself of anything whatsoever about a universe. So, to ground themselves of this half, "they" (female) are not *notalls*, with the consequence and for the same reason, that none of them is all either. (Lacan 1973, pp. 14–15, 22)

Tout peut être maintenu à se développer autour de ce que j'avance de la corrélation logique de deux formules qui, à s'inscrire mathématiquement $\forall x \cdot \Phi x$, et $\exists x \cdot \overline{\Phi x}$, s'énoncent:

la première, pour tout x, Φx est satisfait, ce qui peut se traduire d'un V notant valeur de vérité. Ceci, traduit dans le discours analytique dont c'est la pratique de faire sens, "veut dire" que tout sujet en tant que tel, puisque c'est là l'enjeu de ce discours, s'inscrit dans la fonction phallique pour parer à l'absence du rapport sexuel (la pratique de faire sens, c'est justement de se referer à cet ab-sens);

la seconde, il y a par exception le cas, familier en mathématique (l'argument $x = 0$ dans la fonction exponentielle $1/x$), le cas où il existe un x pour lequel Φx, la fonction, n'est pas satisfaite, c'est-à-dire ne fonctionnant pas, est exclue de fait.

C'est précisément d'où je conjugue le tous de l'universelle, plus modifié qu'on ne s'imagine dans le *pourtout* du quanteur, à l'*il existe un* que le quantique lui apparie, sa différence étant patente avec ce qu'implique la proposition qu'Aristote dit particulière. Je les conjugue de ce que l'*il existe un* en question, à faire limite au *pourtout*, est ce qui l'affirme ou le confirme (ce qu'un proverbe objecte déjà au contradictoire d'Aristote).

. . .

Que j'énonce l'existence d'un sujet à la poser d'un dire que non à la fonction propositionnelle Φx, implique qu'elle s'inscrive d'un quanteur dont cette fonction se trouve coupée de ce qu'elle n'ait en ce point aucune valeur qu'on puisse noter de vérité, ce qui veut dire d'erreur pas plus, le faux seulement à entendre *falsus* comme du chu, ce où j'ai déjà mis l'accent.

En logique classique, qu'on y pense, le faux ne s'aperçoit pas qu'à être de la vérité l'envers, il la désigne aussi bien.

Il est donc juste d'écrire comme je le fais: $\mathrm{E}x \cdot \overline{\Phi x}$.

. . .

De deux modes dépend que le sujet ici se propose d'être dit femme. Les voici:

$$\overline{\mathrm{E}x} \cdot \overline{\Phi x} \text{ et } \overline{\mathrm{A}x} \cdot \Phi x.$$

Leur inscription n'est pas d'usage en mathématique. Nier, comme la barre mise au-dessus du quanteur le marque, nier qu'*existe un* ne se fait pas, et moins encore que *pourtout* se pourpastoute.

C'est là pourtant que se livre le sens du dire, de ce que, s'y conjuguant le *nyania* qui bruit des sexes en compagnie, il supplée à ce qu'entre eux, de rapport nyait pas.

Ce qui est à prendre non pas dans le sens qui, de réduire nos quanteurs à leur lecture selon Aristote, égalerait le *nex-istun* au *nulnest* de son universelle négative, ferait revenir le μή πάντες, le *pastout* (qu'il a pourtant su formuler), à témoigner de l'existence d'un sujet à dire que non à la fonction phallique, ce à le supposer de la contrariété dite de deux particulières.

Ce n'est pas là le sens du dire, qui s'inscrit de ces quanteurs.

Il est: que pour s'introduire comme moitié à dire des femmes, le sujet se détermine de ce que, n'existant pas de suspens à la fonction phallique, tout puisse ici s'en dire, même à provenir du sans raison. Mais c'est un tout d'hors univers, lequel se lit tout de go du second quanteur comme *pastout*.

Le sujet dans la moitié où il se détermine des quanteurs niés, c'est de ce que rien d'existant ne fasse limite de la fonction, que ne saurait s'en assurer quoi que ce soit d'un univers.

Ainsi à se fonder de cette moitié, "elles" ne sont *pastoutes*, avec pour suite et du même fait, qu'aucune non plus n'est toute. (Lacan 1973, pp. 14–15, 22)

Among the other examples of sophisticated terminology thrown at the reader, let us note in Lacan (1971): *union* (in mathematical logic) (p. 206) and *Stokes' theorem* (a particularly shameless case) (p. 213). In Lacan (1975c): *gravitation* ("unconscious of the particle"!) (p. 100). In Lacan (1988): *theory of the unified field* (p. 239). And in Lacan (1998): *Bourbaki* (pp. 28, 47), *quark* (p. 36), *Copernicus* and *Kepler* (pp. 41–43), *inertia*, $mv^2/2$, *mathematical formalization* (p. 130).

Conclusion

What should we make of Lacan's mathematics? Commentators disagree about Lacan's intentions: to what extent was he aiming to "mathematize" psychoanalysis? We are unable to give any definitive answer to this question—which, in any case, does not matter much, since Lacan's "mathematics" are so bizarre that they cannot play a fruitful role in any serious psychological analysis.

To be sure, Lacan does have a vague idea of the mathematics he invokes (but not much more). It is not from him that a student will learn what a natural number or a compact set is, but his statements, when they are understandable, are not always false. On the other hand, he excels (if we may use this word) at the second type of abuse listed in our introduction: his analogies between psychoanalysis and mathematics are the most arbitrary imaginable, and he gives absolutely no empirical or conceptual justification for them (neither here nor elsewhere in his work). Finally, as for showing off a superficial erudition and manipulating meaningless sentences, the texts quoted above surely speak for themselves.

Let us conclude with some general observations about Lacan's oeuvre. We emphasize that these remarks go far beyond what we can claim to have proven in this chapter, and so should be regarded as plausible conjectures meriting more detailed study.

The most striking aspect of Lacan and his disciples is probably their attitude towards science, and the extreme privilege they accord to "theory" (in actual fact, to formalism and wordplay) at the expense of observations and experiments. After all, psychoanalysis, assuming that it has a scientific basis, is a rather young science. Before launching into vast theoretical generalizations, it might be prudent to check the empirical adequacy of at least some of its propositions. But, in Lacan's writings, one finds mainly quotations and analyses of texts and concepts.

Lacan's defenders (as well as those of the other authors discussed here) tend to respond to these criticisms by resorting to a strategy that we shall call "neither/nor": these writings should be evaluated neither as science, nor as philosophy, nor as poetry, nor . . . One is then faced with what could be called a "secular mysticism": mysticism because the discourse aims at producing mental effects that are not purely aesthetic, but without addressing itself to reason; secular because the cultural references (Kant, Hegel, Marx, Freud, mathematics, contemporary literature . . .) have nothing to do with traditional religions and are attractive to the modern reader. Furthermore, Lacan's writings became, over time, increasingly cryptic—a characteristic common to many sacred texts—by combining plays on words with fractured syntax; and they served as a basis for the reverent exegesis undertaken by his disciples. One may then wonder whether we are not, after all, dealing with a new religion.

3. Julia Kristeva

Julia Kristeva changes the order of things; she always destroys the latest preconception, the one we thought we could be comforted by, the one of which we could be proud: what she displaces is the already-said, that is to say, the insistence of the signified, that is to say, silliness; what she subverts is the authority of monologic science and of filiation. Her work is entirely new and precise . . .

—*Roland Barthes (1970, p. 19), concerning Kristeva's Séméiotiké: Researches for a Semioanalysis*

The works of Julia Kristeva touch on many areas, from literary criticism to psychoanalysis to political philosophy. We shall analyze here some excerpts from her early work on linguistics and semiotics. These texts, which date from the late 1960s to the mid-1970s, cannot properly be called poststructuralist; they belong, rather, to the worst excesses of structuralism. Kristeva's declared goal is to construct a formal theory of poetic language. This goal is, however, ambiguous because, on the one hand, she asserts that poetic language is "a formal system whose theorization can be based on [mathematical] set theory", and on the other hand, she says in a footnote that this is "only metaphorical".

Metaphor or not, this enterprise faces a serious problem: What relation, if any, does poetic language have with mathematical set theory? Kristeva doesn't really say. She invokes technical notions concerning infinite sets, whose relevance to poetic language is difficult to fathom, especially since no argument is given. Moreover, her presentation of the mathematics contains some gross errors, for example with regard to Gödel's theorem. Let us emphasize that Kristeva has long since abandoned this approach; nevertheless, it is too typical of the type of work we are criticizing for us to pass it over in silence.

The excerpts below come mainly from Kristeva's celebrated book *Séméiotiké: Researches for a Semioanalysis* (1969).[36] One of her interpreters describes this work as follows:

> What is most striking about Kristeva's work . . . is the competence with which it is presented, the intense single-mindedness with which it is pursued, and finally, its intricate rigour. No resources are spared: existing theories of logic are invoked and, at one point, quantum mechanics . . . (Lechte 1990, p. 109)

Let us therefore examine some examples of this competence and rigor:

> . . . science is a logical endeavor based on the Greek (Indo-European) sentence that is constructed as subject-predicate and that proceeds by identification, determination, causality.[37] Modern logic from Frege and Peano through Lukasiewicz, Ackermann or Church, which moves in the dimensions 0–1, and even Boole's logic which, starting from set theory, gives formalizations that are more isomorphic to the functioning of language, are inoperative in the sphere of poetic language where 1 is not a limit.

---

[36]One of Kristeva's commentators, Toril Moi, explains the context:

In 1966 Paris witnessed not only the publication of Jacques Lacan's *Écrits* and Michel Foucault's *Les Mots et les choses*, but also the arrival of a young linguist from Bulgaria. At the age of 25, Julia Kristeva . . . took the Left Bank by storm. . . . Kristeva's linguistic research was soon to lead to the publication of two important books, *Le Texte du roman* and *Séméiotiké*, and to culminate with the publication of her massive doctoral thesis, *La Révolution du langage poétique*, in 1974. This theoretical production earned her a chair in linguistics at the University of Paris VII. (Moi 1986, p. 1)

[37]Here Kristeva seems to be appealing implicitly to the "Sapir-Whorf thesis" in linguistics, that is, *grosso modo*, the idea that our language radically conditions our view of the world. This thesis is nowadays sharply criticized by some linguists: see, for example, Pinker (1995, pp. 57–67).

It is therefore impossible to formalize poetic language using the existing logical (scientific) procedures without denaturing it. A literary semiotics has to be made starting from a *poetic logic*, in which the concept of *power of the continuum*[38] would encompass the interval from 0 to 2, a continuum where 0 denotes and 1 is implicitly transgressed. (Kristeva 1969, pp. 150–151, italics in the original[39])

In this excerpt, Kristeva makes one correct assertion and two mistakes. The correct assertion is that poetic sentences cannot, in general, be evaluated as true or false. Now, in mathematical logic, the symbols 0 and 1 are used to denote "false" and "true", respectively; it is in this sense that Boole's logic uses the set {0,1}. Kristeva's allusion to mathematical logic is thus correct, though it adds nothing to the initial observation. But in the second paragraph, she seems to confuse the *set* {0,1}, which is composed of the two elements 0 and 1, with the *interval* [0,1], which contains all the real numbers between 0 and 1. The latter set, unlike the former, is an *infinite* set, which, moreover, has the power of the continuum (see note 38). Besides, Kristeva lays great stress on the fact that she has a set (the interval from 0 to 2) that "transgresses" 1, but from the point of view she purports to adopt—that of the cardinality (or power) of sets—there is

---

[38]The "power of the continuum" is a concept belonging to the mathematical theory of infinite sets, which was developed by Georg Cantor and other mathematicians starting in the 1870s. It turns out that there are many different "sizes" (or *cardinalities*) of infinite sets. Some infinite sets are termed *countable* (or *denumerable*): for example, the set of all positive integers (1, 2, 3, . . .) or, more generally, any set whose elements can be put into one-to-one correspondence with the set of all positive integers. On the other hand, Cantor proved in 1873 that there does *not* exist a one-to-one correspondence between the integers and the set of all *real* numbers. Therefore, the real numbers are in a certain sense "more numerous" than the integers: they are said to have the *cardinality (or power) of the continuum*, as do all those sets that can be put in one-to-one correspondence with them. Let us remark (what is at first surprising) that one can establish a one-to-one correspondence between the real numbers and the real numbers contained in an interval: for example, those numbers between 0 and 1, or those between 0 and 2, etc. More generally, every infinite set can be put into one-to-one correspondence with some of its proper subsets.

[39]Translation ours. A slightly different translation of this excerpt and the next one can be found in Kristeva (1980, pp. 70–72).

no difference between the interval [0,1] and the interval [0,2]: both have the power of the continuum.

In the subsequent text, these two errors become even more manifest:

> In this "power of the continuum" from zero to the specifically poetic double, one notices that the linguistic, psychic, and social "prohibition" [*interdit*] is 1 (God, the law, the definition), and that the only linguistic practice that "escapes" from this prohibition is poetic discourse. It is no accident that the inadequacies of Aristotelian logic in its application to language were pointed out, on the one hand, by the Chinese philosopher Chang Tung-sun who came from another linguistic realm (that of ideograms) where the Yin-Yang "dialogue" is deployed in place of God, and on the other hand, by Bakhtin who attempted to go beyond the Formalists by a dynamic theorization carried out in a revolutionary society. For him, narrative discourse, which he assimilates to epic discourse, is a prohibition, a *"monologism"*, a subordination of the code to 1, to God. Consequently, the epic is religious and theological, and any "realist" narration obeying the 0–1 logic is dogmatic. The realist novel that Bakhtin calls monologic (Tolstoy) tends to evolve in that space. Realist description, the definition of a "personality type" [*caractère*], the creation of a "character" [*personnage*], the development of a "subject": all these descriptive elements of narrative belong to the interval 0–1 and thus are *monologic*. The only discourse in which the 0–2 poetic logic is fully realized would be that of the carnival: it transgresses the rules of the linguistic code, as well as that of social morality, by adopting a dream-like logic.
>
> ... A new approach to poetic texts can be sketched starting from this term [dialogism] that literary semiotics can adopt. The logic implied by "dialogism" is simultaneously: ...
> 3) A logic of the *"transfinite"*[40], a concept that we borrow

---

[40] In mathematics, the word "transfinite" is more or less synonymous with "infinite". It is used most commonly to characterize a "cardinal number" or an "ordinal number".

from Cantor, which introduces, starting from the "power of the continuum" of poetic language (0–2), a second formative principle, namely: a poetic sequence is "next-larger" (not causally deduced) to all the preceding sequences of the Aristotelian series (scientific, monologic, narrative). Then, the ambivalent space of the novel presents itself as ordered by two formative principles: the monologic (each successive sequence is determined by the preceding one) and the dialogic (transfinite sequences that are next-larger to the preceding causal sequence). [**Footnote:** Let us emphasize that the introduction of notions from set theory in an analysis of poetic language is only metaphorical: it is possible because an analogy can be established between the relations Aristotelian logic/poetic logic on the one hand, and denumerable/infinite on the other.] (Kristeva 1969, pp. 151–153, italics in the original)

At the end of this passage, Kristeva concedes that her "theory" is only a metaphor. But even at that level, she provides no justification: far from having established an analogy between "Aristotelian logic/poetic logic" and "denumerable/infinite", she has merely invoked the *names* of these latter concepts, without giving the slightest explanation of their *meaning* or, above all, their relevance (even metaphorical) for "poetic logic". For what it's worth, the theory of transfinite numbers has nothing to do with causal deduction.

Later on in the text, Kristeva returns to mathematical logic:

For us poetic language is not a code encompassing the others, but a class A that has the same power as the function $\varphi(x_1 \ldots x_n)$ of the infinity of the linguistic code (see the existence theorem, cf. p. 189), and all the "other languages" (the "usual" language, the "meta-languages", etc.) are quotients of A over more restricted extents [*étendues*] (limited by the rules of the subject-predicate construction, for example, as being at the basis of formal logic), and disguising, because of this limitation, the morphology of the function $\varphi(x_1 \ldots x_n)$.

Poetic language (which we shall henceforth denote by the initials pl) contains the code of linear logic. Moreover, we can find in it all the combinatoric figures that algebra has formalized in a system of artificial signs and that are not externalized at the level of the manifestation of the usual language. . . .

The pl cannot, therefore, be a sub-code. It is the infinite ordered code, a complementary system of codes from which one may isolate (by operatory abstraction and by way of proof of a theorem) a usual language, a scientific metalanguage and all the artificial systems of signs—which are all only subsets of this infinite, externalizing the rules of its order over a restricted space (their power is lesser relative to that of the pl that is surjected onto them). (Kristeva 1969, pp. 178–179)

These paragraphs are meaningless, though Kristeva has very ably strung together a series of mathematical terms. But it gets even better:

Having assumed that poetic language is a formal system whose theorization can be based on *set theory*, we may observe, at the *same time*, that the functioning of poetic meaning obeys the principles designated by the *axiom of choice*. This axiom specifies that there exists a single-valued correspondence, represented by a class, which associates to each non-empty set of the theory (of the system) one of its elements:

$$(\exists A)\, \{Un(A) \cdot (x)[\sim\!Em(x)\cdot \supset \cdot (\exists y)[y \in x \cdot \langle yx \rangle \in A]]\}$$

[$Un(A)$ — "A is single-valued"; $Em(x)$ — "the class x is empty".]

Said otherwise, one can choose simultaneously an element in each of the non-empty sets that we consider. So stated, the axiom is applicable in our universe E of the *pl*. It makes precise how every sequence contains the message of the book. (Kristeva 1969, p. 189, italics in the original)

These paragraphs (as well as the following ones) illustrate brilliantly the acerbic comments of the sociologist Stanislav Andreski quoted in our Introduction (p. 11). Kristeva never explains the relevance of the axiom of choice for linguistics (in our opinion it has none). The axiom of choice says that if we have a collection of sets, each of which contains at least one element, then there exists a set containing exactly one element "chosen" from each of the original sets. This axiom permits one to assert the existence of certain sets without constructing them explicitly (one does not say how the "choice" is made). The introduction of this axiom in mathematical set theory is motivated by the study of infinite sets, or of infinite collections of sets. Where does one find such sets in poetry? To say that the axiom of choice "makes precise how every sequence contains the message of the book" is ludicrous—we're unsure whether this assertion does more violence to mathematics or to literature.

Nevertheless, Kristeva continues:

> The compatibility of the axiom of choice and the generalized continuum hypothesis[41] with the axioms of set theory places us at the level of reasoning about the theory, thus in a *metatheory* (and such is the status of semiotic reasoning) whose metatheorems have been perfected [*mis au point*] by Gödel. (Kristeva 1969, p. 189, italics in the original)

Here again, Kristeva is trying to impress the reader with technical jargon. She has indeed cited some very important (meta)-theorems of mathematical logic, but without bothering to ex-

---

[41]As we saw in note 38 above, there exist infinite sets of different "sizes" (called *cardinals*). The smallest infinite cardinal, called "countable" (or "denumerable"), is the one corresponding to the set of all positive integers. A larger cardinal, called the "cardinal of the continuum", is the one corresponding to the set of all real numbers. The continuum hypothesis (CH), introduced by Cantor in the late nineteenth century, asserts that there is no "intermediate" cardinal between the countable and the continuum. The generalized continuum hypothesis (GCH) is an extension of this idea to vastly larger infinite sets. In 1964, Cohen proved that the CH (as well as the GCH) is independent of the other axioms of set theory, in the sense that neither it nor its negation is provable using those axioms.

plain to the reader the *content* of these theorems, much less their relevance for linguistics. (Let us note that the set of all texts ever written, in the entirety of human history, is a *finite* set. Moreover, any natural language—for example, English or Chinese—has a finite alphabet; a sentence, or even a book, is a finite sequence of letters. Therefore, even the set of *all* finite sequences of letters in *all* conceivable books, without any restriction on their length, is a *denumerable* infinite set. It is hard to see how the continuum hypothesis, which concerns nondenumerable infinite sets, could have any application in linguistics.)

All this does not prevent Kristeva from pushing onward:

One finds there precisely the *existence theorems* that we do not intend to develop here, but that interest us to the extent that they provide *concepts* allowing us to pose in a new way— a way that would be impossible without them—the *object* that interests us: poetic language. The generalized existence theorem postulates, as one knows, that:

"If $\varphi(x_1, \ldots, x_n)$ is a primitive propositional function containing no free variables other than $x_1, \ldots, x_n$, without necessarily containing all of them, there exists a class A such that, for all *sets* $x_1, \ldots, x_n$, $\langle x_1, \ldots, x_n \rangle \in A \cdot \equiv \cdot \varphi(x_1, \ldots, x_n)$."[42]

In the poetic language, this theorem denotes the different sequences as equivalent to a function encompassing all of them. Two consequences follow from this: 1) it stipulates the non-causal chaining [*enchaînement*] of poetic language and the expansion of the letter in the book; 2) it stresses the range [*portée*] of this literature which puts forth its message in the smallest sequences: the meaning (φ) is contained in the mode of junction of words, of sentences . . .

---

[42]This is a technical result of Gödel-Bernays set theory (one of the versions of axiomatic set theory). Kristeva does not explain its relevance for poetic language. Let us note in passing that to precede such a technical statement by the expression "as one knows" *(on le sait)* is a typical example of intellectual terrorism.

Lautréamont was one of the first to consciously practice this theorem.[43]

The notion of constructibility implied by the axiom of choice associated to what we have just set forth for poetic language, explains the impossibility of establishing a contradiction in the space of poetic language. This observation is close to Gödel's observation concerning the impossibility of proving the inconsistency [*contradiction*] of a system by means formalized within the system. (Kristeva 1969, pp. 189–190, italics in the original)

In this excerpt, Kristeva shows that she does not understand the mathematical concepts she invokes. First of all, the axiom of choice does not imply any "notion of constructibility"; quite the contrary, it allows one to assert the existence of some sets *without* having a rule to "construct" them (see above). Secondly, Gödel proved exactly the opposite of what Kristeva claims, namely the impossibility of establishing, by means formalizable within the system, the system's *consistency* (i.e. *non-contradiction*).[44]

Kristeva has also tried to apply set theory to political philosophy. The following excerpt is taken from her book *Revolution in Poetic Language* (1974):

---

[43]It is rather improbable that Lautréamont (1846–1870) could have "consciously practiced" a theorem of Gödel-Bernays set theory (developed between 1937 and 1940) or even of set theory *tout court* (developed after 1870 by Cantor and others).

[44]Gödel, in his famous article (1931), proved two principal theorems concerning the incompleteness of certain formal systems (complex enough to encode elementary arithmetic) in mathematical logic. Gödel's first theorem exhibits a proposition that is neither provable nor refutable in the given formal system, provided that this system is consistent. (One may nevertheless see, using reasoning that cannot be formalized within the system, that this proposition is *true*.) Gödel's second theorem asserts that, if the system is consistent, it is impossible to prove this property by means that can be formalized within the system itself.

On the other hand, it is very easy to invent inconsistent (i.e. self-contradictory) systems of axioms; and, when a system is inconsistent, there always exists a proof of this inconsistency by means formalized within the system: although this proof may sometimes be difficult to find, it exists, almost by virtue of the definition of the word "inconsistent".

For an excellent introduction to Gödel's theorem, see Nagel and Newman (1958).

A discovery of Marx, which has not heretofore been sufficiently emphasized, can be sketched here. If each individual
or each social organism represents a set, the set of all sets that
the State should be does not exist. The State as set of all sets
is a fiction, it cannot exist, just as there does not exist a set of
all sets in set theory.[45] [**Footnote:** On this topic, cf. Bourbaki[46], but also, concerning the relations between set theory
and the functioning of the unconscious, D. Sibony, "Infinity
and castration", in *Scilicet*, No. 4, 1973, pp. 75–133.] The State
is, at most, a collection of all the finite sets. But for this collection to exist, and for finite sets to exist too, there must be
some infinity: the two propositions are equivalent. The desire
to form the set of all finite sets puts the infinite on stage, and
reciprocally. Marx, who noticed the illusion of the State to be
the set of all sets, saw in the social unit as presented by the
bourgeois Republic a collection that nevertheless constitutes,
for itself, a set (just as the collection of the finite ordinals is a
set if one poses it as such) from which something is lacking:
indeed, its *existence* or, if one wants, its *power* is dependent
on the existence of the infinite that no other set can contain.
(Kristeva 1974, pp. 379–380, italics in the original)

But Kristeva's mathematical erudition is not limited to set
theory. In her article "On the subject in linguistics", she applies
mathematical analysis and topology to psychoanalysis:

[I]n the syntactic operations following the mirror stage, the
subject is already sure of his uniqueness: his flight towards the
"point ∞" in the signifying [*signifiance*] is stopped. One
thinks for example of a set C_0 on a usual space R^3 where for
every continuous function F on R^3 and each integer $n > 0$, the

---

[45]See note 31 above. It must be emphasized that no problem occurs for finite sets, such as the set of individuals in a society.

[46]Nicolas Bourbaki is the pseudonym of a group of prominent French mathematicians who, since the late 1930s, have published about thirty volumes of their series, *Elements of Mathematics*. But, despite the title, these books are far from elementary. Whether or not Kristeva has read Bourbaki, this reference has no function other than to impress the reader.

set of points X where $F(X)$ exceeds n is *bounded*, the functions of C_0 tending to 0 when the variable X recedes towards
the "other scene". In this topos, the subject placed in C_0 does
not reach this "center exterior to language" about which
Lacan speaks and where he loses himself as subject, a situation that would translate the relational group that topology
calls a *ring*. (Kristeva 1977, p. 313, italics in the original)

This is one of the best examples of Kristeva's attempts to impress the reader with fancy words that she obviously does not
understand. Andreski "advised" the budding social scientist to
copy the less complicated parts of a mathematics textbook; but
the definition given here of the set of functions $C_0(\mathrm{R}^3)$ is not
even correctly copied, and the errors stand out to anyone who
understands the subject.[47] But the real problem is that the purported application to psychoanalysis is nonsense. How could a
"subject" be "placed in C_0"?

Among the other examples of mathematical terminology that
Kristeva uses without explanation or justification, let us note in
Kristeva (1969): *stochastic analysis* (p. 177), *Hilbert's finitism*
(p. 180), *topological space* and *abelian ring* (p. 192), *union*
(p. 197), *idempotence, commutativity, distributivity, . . .* (pp.
258–264), *Dedekind structure with orthocomplements* (pp.
265–266), *infinite functional Hilbert spaces* (p. 267), *algebraic
geometry* (p. 296), *differential calculus* (pp. 297–8). And in Kristeva (1977): *articulation set* in graph theory (p. 291), *predicate
logic* (which she bizarrely calls "modern proportional logic"[48])
(p. 327).

---

[47]The space $C_0(\mathrm{R}^3)$ is composed of all the real-valued continuous functions, defined
on R^3, that "tend to zero at infinity". But, in the precise definition of this concept,
Kristeva should have said: (a) $|F(X)|$ instead of $F(X)$; (b) "exceeds $1/n$" instead of
"exceeds n"; and (c) "containing all the continuous functions F on R^3 such that"
instead of "where for every continuous function F on R^3".

[48]This malapropism probably arises from a combination of two mistakes: on the one
hand, it seems that Kristeva has confused predicate logic with propositional logic;
and on the other hand, she or her editors have apparently inserted the typographical
error "proportional" *(proportionnelle)* in place of "propositional" *(propositionnelle)*.

To summarize, our evaluation of Kristeva's scientific abuses is similar to the one we gave for Lacan. In general, Kristeva has at least a vague idea of the mathematics she invokes, even if she manifestly does not always understand the meaning of the words she uses. But the main problem raised by these texts is that she makes no effort to justify the *relevance* of these mathematical concepts to the fields she is purporting to study—linguistics, literary criticism, political philosophy, psychoanalysis—and this, in our opinion, is for the very good reason that there is none. Her sentences are more meaningful than those of Lacan, but she surpasses even him for the superficiality of her erudition.

4. Intermezzo: Epistemic Relativism in the Philosophy of Science

I did not write this work merely with the aim of setting the exegetical record straight. My larger target is those contemporaries who—in repeated acts of wish-fulfillment—have appropriated conclusions from the philosophy of science and put them to work in aid of a variety of social cum political causes for which those conclusions are ill adapted. Feminists, religious apologists (including "creation scientists"), counterculturalists, neo-conservatives, and a host of other curious fellow-travelers have claimed to find crucial grist for their mills in, for instance, the avowed incommensurability and underdetermination of scientific theories. The displacement of the idea that facts and evidence matter by the idea that everything boils down to subjective interests and perspectives is—second only to American political campaigns—the most prominent and pernicious manifestation of anti-intellectualism in our time.

—*Larry Laudan, Science and Relativism (1990, p. x)*

Since much postmodern discourse flirts with one form or another of cognitive relativism or invokes arguments that can support it, it seems useful at this point to include an epistemological discussion. We are aware that we will be dealing with difficult problems concerning the nature of knowledge and objectivity, which have worried philosophers for centuries. It is not necessary to share our philosophical positions in order to agree with the rest of what we say. In this chapter we shall criticize ideas that are in our view erroneous, but which are sometimes (not always) so for subtle reasons, contrary to the texts we criticize in the rest of this book. Our philosophical argumentation will, in any case, be rather minimalist; we shall not enter into the more delicate philosophical debates between, for example, moderate forms of realism and instrumentalism.

We are concerned here with a potpourri of ideas, often poorly formulated, that go under the generic name of "relativism" and are nowadays rather influential in some sectors of the academic humanities and social sciences. This relativist zeitgeist originates partly from contemporary works in the philosophy of science, such as Thomas Kuhn's *The Structure of Scientific Revolutions* and Paul Feyerabend's *Against Method*, and partly from extrapolations of these philosophers' work by their successors.[49] Of course, we do not purport to examine the entire work of the authors discussed in this chapter; that would be an unmanageable task. Rather, we shall limit ourselves to an analysis of certain texts that illustrate rather widespread ideas. We shall show that these texts are often ambiguous and can be read in at least two distinct ways: a "moderate" reading, which leads to claims that are either worth discussing or else true but trivial; and a "radical" reading, which leads to claims that are surprising but false. Unfortunately, the radical interpretation is often taken not only as the "correct" interpretation of the original text but also as a well-established fact ("X has shown that . . .")—a conclusion that we shall sharply criticize. It might, of course, be argued that no one holds this radical interpretation; and all the better if that is true. But the numerous discussions we have had during which the theory-ladenness of observation, the underdetermination of theory by evidence or the alleged incommensurability of paradigms have been put forward in order to support relativist positions leave us rather skeptical. And to show that we are not criticizing a figment of our imagination, we shall give, at the end of this chapter, a few practical examples of the relativism that is widespread in the United States, in Europe, and in parts of the Third World.

Roughly speaking, we shall use the term "relativism" to designate any philosophy that claims that the truth or falsity of a statement is relative to an individual or to a social group. One may distinguish different forms of relativism according to the

---

[49] There are, of course, many other sources of the relativist zeitgeist, from Romanticism to Heidegger, but we shall not deal with them here.

nature of the statement in question: *cognitive* or *epistemic* relativism when one is dealing with an assertion of fact (that is,
about what exists or is claimed to exist); *moral* or *ethical* relativism when one is dealing with a value judgment (about what
is good or bad, desirable or pernicious); and *aesthetic* relativism
when one is dealing with an artistic judgment (about what is
beautiful or ugly, pleasant or unpleasant). Here we shall be concerned only with *epistemic* relativism and not with moral or
aesthetic relativism, which raise very different issues.

We are well aware that we will be criticized for our lack of
formal philosophical training. In the Introduction we explained
why this sort of objection leaves us cold, but it seems particularly irrelevant here. After all, there is no doubt that the relativist
attitude is at odds with scientists' idea of their own practice.
While scientists try, as best they can, to obtain an objective view
of (certain aspects of) the world[50], relativist thinkers tell them
that they are wasting their time and that such an enterprise is,
in principle, an illusion. We are thus dealing with a fundamental
conflict. And as physicists who have long pondered the foundations of our discipline and of scientific knowledge in general,
we think it important to try to give a reasoned answer to the relativist objections, even though neither of us holds a diploma in
philosophy.

We shall start by sketching our attitude toward scientific
knowledge[51], and shall then review briefly some aspects of
twentieth-century epistemology (Popper, Quine, Kuhn, Feyerabend); our aim will mostly be to disentangle some of the confusions concerning notions such as "underdetermination" and

---

[50]With, of course, many nuances about the meaning of the word "objective", which are
reflected, for instance, in the opposition between such doctrines as realism,
conventionalism and positivism. Nevertheless, few scientists would be ready to admit
that the whole of scientific discourse is a mere social construction. As one of us
wrote, we have no desire to be the Emily Post of quantum field theory (Sokal 1996c,
p. 94, reproduced here in Appendix C).

[51]Limiting ourselves to the natural sciences and taking most of the examples from our
own field, physics. We shall not deal with the delicate question of the scientificity of
the various social sciences.

"incommensurability". Finally, we shall examine critically some recent tendencies in the sociology of science (Barnes, Bloor, Latour) and shall give some practical examples of the effects of contemporary relativism.

Solipsism and Radical Skepticism

> When my brain excites in my soul the sensation of a tree, or of a house, I pronounce, without hesitation, that a tree, or a house, really exists out of me, of which I know the place, the size, and other properties. Accordingly, we find neither man nor beast who calls this truth in question. If a peasant should take it into his head to conceive such a doubt, and should say, for example, he does not believe that his bailiff exists, though he stands in his presence, he would be taken for a madman, and with good reason; but when a philosopher advances such sentiments, he expects we should admire his knowledge and sagacity, which infinitely surpass the apprehensions of the vulgar.
> —*Leonhard Euler (1997 [1761], pp. 428–429)*

Let us start at the beginning. How can one possibly hope to attain an objective (albeit approximate and incomplete) knowledge of the world? We never have direct access to the world; we have direct access only to our sensations. How do we know that there even *exists* anything outside of those sensations?

The answer, of course, is that we have no *proof;* it is simply a perfectly reasonable hypothesis. The most natural way to explain the persistence of our sensations (in particular, the unpleasant ones) is to suppose that they are caused by agents outside our consciousness. We can almost always change at will the sensations that are pure products of our imagination, but we cannot stop a war, stave off a lion, or start a broken-down car by pure thought alone. Nevertheless—and it is important to emphasize this—this argument *does not refute* solipsism. If anyone insists that he is a "harpsichord playing solo" (Diderot), there is no way to convince him of his error. However, we have never met a sincere solipsist and we doubt that any exist.[52] This illustrates an important principle that we shall use several times in this chapter: *the mere fact that an idea is irrefutable does not imply that there is any reason to believe it is true.*

Another position that one sometimes encounters, in place of solipsism, is radical skepticism: "Of course there exists an external world, but it is impossible for me to obtain any reliable knowledge of that world." In essence the argument is the same as that of the solipsist: I have immediate access only to my sensations; how can I know whether they *accurately reflect* reality? To be certain that they do, I would need to invoke an *a priori* argument, such as the proof of the existence of a benevolent deity in Descartes' philosophy; and such arguments have fallen into disfavor in modern philosophy, for all sorts of good reasons that we need not rehearse here.

This problem, like many others, was very well formulated by Hume:

> It is a question of fact, whether the perceptions of the senses be produced by external objects, resembling them: how shall this question be determined? By experience surely; as all other questions of a like nature. But here experience is, and must be entirely silent. The mind has never anything present to it but the perceptions, and cannot possibly reach any experience of their connexion with objects. The supposition of such a connexion is, therefore, without any foundation in reasoning. (Hume 1988 [1748], p. 138: *An Enquiry Concerning Human Understanding*, Section XII, Part I)

What attitude should one adopt toward radical skepticism? The key observation is that such skepticism applies to *all* our knowledge: not only to the existence of atoms, electrons or genes, but also to fact that blood circulates in our veins, that the

---

[52]Bertrand Russell (1948, p. 196) tells the following amusing story: "I once received a letter from an eminent logician, Mrs Christine Ladd Franklin, saying that she was a solipsist, and was surprised that there were not others". We learned this reference from Devitt (1997, p. 64).

Earth is (approximately) round, and that at birth we emerged from our mother's womb. Indeed, even the most commonplace knowledge in our everyday lives—there is a glass of water in front of me on the table—depends entirely on the supposition that our perceptions do not *systematically* mislead us and that they are indeed produced by external objects that, in some way, resemble those perceptions.[53]

The universality of Humean skepticism is also its weakness. Of course, it is irrefutable. But since no one is systematically skeptical (when he or she is sincere) with respect to ordinary knowledge, one ought to ask *why* skepticism is rejected in that domain and *why* it would nevertheless be valid when applied elsewhere, for instance, to scientific knowledge. Now, the reason why we reject systematic skepticism in everyday life is more or less obvious and is similar to the reason we reject solipsism. The best way to account for the coherence of our experience is to suppose that the outside world corresponds, at least approximately, to the image of it provided by our senses.[54]

Science As Practice

> For my part, I have no doubt that, although progressive changes are to be expected in physics, the present doctrines are likely to be nearer to the truth than any rival doctrines now before the world. Science is at no moment quite right, but it is seldom quite wrong, and has, as a rule, a better chance of being right than the theories of the unscientific. It is, therefore, rational to accept it hypothetically.
>
> —*Bertrand Russell, My Philosophical Development*
> *(1995 [1959], p. 13)*

---

[53] To claim this does not mean that we claim to have an entirely satisfactory answer to the question of *how* such a correspondence between objects and perceptions is established.

[54] This hypothesis receives a deeper explanation with the subsequent development of science, in particular of the biological theory of evolution. Clearly, the possession of sensory organs that reflect more or less *faithfully* the outside world (or, at least, some important aspects of it) confers an evolutionary advantage. Let us stress that this argument does not refute radical skepticism, but it does increase the coherence of the anti-skeptical worldview.

Once the general problems of solipsism and radical skepticism have been set aside, we can get down to work. Let us suppose that we are able to obtain some more-or-less reliable knowledge of the world, at least in everyday life. We can then ask: *To what extent* are our senses reliable or not? To answer this question, we can compare sense impressions among themselves and vary certain parameters of our everyday experience. We can map out in this way, step by step, a practical rationality. When this is done systematically and with sufficient precision, science can begin.

For us, the scientific method is not radically different from the rational attitude in everyday life or in other domains of human knowledge. Historians, detectives, and plumbers—indeed, all human beings—use the same basic methods of induction, deduction, and assessment of evidence as do physicists or biochemists. Modern science tries to carry out these operations in a more careful and systematic way, by using controls and statistical tests, insisting on replication, and so forth. Moreover, scientific measurements are often much more precise than everyday observations; they allow us to discover hitherto unknown phenomena; and they often conflict with "common sense". But the conflict is at the level of conclusions, not the basic approach.[55,56]

---

[55]For example: Water appears to us as a continuous fluid, but chemical and physical experiments teach us that it is made of atoms.

[56]Throughout this chapter, we stress the methodological continuity between scientific knowledge and everyday knowledge. This is, in our view, the proper way to respond to various skeptical challenges and to dispel the confusions generated by radical interpretations of correct philosophical ideas such as the underdetermination of theories by data. But it would be naive to push this connection too far. Science—particularly fundamental physics—introduces concepts that are hard to grasp intuitively or to connect directly to common-sense notions. (For example: forces acting instantaneously throughout the universe in Newtonian mechanics, electromagnetic fields "vibrating" in vacuum in Maxwell's theory, curved space-time in Einstein's general relativity.) And it is in discussions about the meaning of these theoretical concepts that various brands of realists and anti-realists (e.g., intrumentalists, pragmatists) tend to part company. Relativists sometimes tend to fall back on instrumentalist positions when challenged, but there is a profound difference between the two attitudes. Instrumentalists may want to claim either that we have no way of knowing whether "unobservable" theoretical entities really exist, or that their meaning is defined solely through measurable quantities; but this does not imply that they regard such entities as "subjective" in the sense that their meaning would be

The main reason for believing scientific theories (at least the best-verified ones) is that they explain the coherence of our experience. Let us be precise: here "experience" refers to *all* our observations, including the results of laboratory experiments whose goal is to test quantitatively (sometimes to incredible precision) the predictions of scientific theories. To cite just one example, quantum electrodynamics predicts that the magnetic moment of the electron has the value[57]

$$1.001\ 159\ 652\ 201 \pm 0.000\ 000\ 000\ 030 \ ,$$

where the "±" denotes the uncertainties in the theoretical computation (which involves several approximations). A recent experiment gives the result

$$1.001\ 159\ 652\ 188 \pm 0.000\ 000\ 000\ 004 \ ,$$

where the "±" denotes the experimental uncertainties.[58] This agreement between theory and experiment, when combined with thousands of other similar though less spectacular ones, would be a miracle if science said nothing true—or at least *approximately* true—about the world. The experimental confirmations of the best-established scientific theories, taken together, are evidence that we really have acquired an objective (albeit approximate and incomplete) knowledge of the natural world.[59]

---

significantly influenced by extra-scientific factors (such as the personality of the individual scientist or the social characteristics of the group to which she belongs). Indeed, instrumentalists may regard our scientific theories as, quite simply, the most satisfactory way that the human mind, with its inherent biological limitations, is capable of understanding the world.

[57]Expressed in a well-defined unit which is unimportant for the present discussion.

[58]See Kinoshita (1995) for the theory, and Van Dyck *et al.* (1987) for the experiment. Crane (1968) provides a non-technical introduction to this problem.

[59]Subject, of course, to many nuances on the precise meaning of the phrases "approximately true" and "objective knowledge of the natural world", which are reflected in the diverse versions of realism and anti-realism (see note 56 above). For these debates, see for example Leplin (1984).

Having reached this point in the discussion, the radical skeptic or relativist will ask what distinguishes science from other types of discourse about reality—religions or myths, for example, or pseudo-sciences such as astrology—and, above all, what *criteria* are used to make such a distinction. Our answer is nuanced. First of all, there are some general (but basically negative) epistemological principles, which go back at least to the seventeenth century: to be skeptical of *a priori* arguments, revelation, sacred texts, and arguments from authority. Moreover, the experience accumulated during three centuries of scientific practice has given us a series of more-or-less general methodological principles—for example, to replicate experiments, to use controls, to test medicines in double-blind protocols—that can be justified by rational arguments. However, we do not claim that these principles can be codified in a definitive way, nor that the list is exhaustive. In other words, there does not exist (at least at present) a complete codification of scientific rationality, and we seriously doubt that one could ever exist. After all, the future is inherently unpredictable; rationality is always an adaptation to a new situation. Nevertheless—and this is the main difference between us and the radical skeptics—we think that well-developed scientific theories are in general supported by good arguments, but the rationality of those arguments must be analyzed case-by-case.[60]

To illustrate this, let us consider an example that is in a certain sense intermediate between scientific and ordinary knowledge, namely criminal investigations.[61] There are some cases in which even the hardiest skeptic will find it difficult to doubt, in

---

[60]It is also by proceeding on a case-by-case basis that one can appreciate the immensity of the gulf separating the sciences from the pseudo-sciences.

[61]We hasten to add—as if this should even be necessary—that we harbor no illusions about the behavior of real-life police forces, which are by no means always and exclusively dedicated to finding the truth. We employ this example solely to illustrate the abstract epistemological question in a simple concrete context, namely: Suppose that one *does* wish to find the truth about a practical matter (such as who committed a murder); how would one go about it? For an extreme example of this misreading—in which we are compared to former Los Angeles Detective Mark Fuhrman (of O.J. Simpson fame) and his infamous Brooklyn counterparts—see Robbins (1998).

practice, that the culprit has really been found: one may, after all, possess the weapon, fingerprints, DNA evidence, documents, a motive, and so forth. Nevertheless, the path leading to those discoveries can be very complicated: the investigator has to make decisions (on the leads to follow, on the evidence to seek) and draw provisional inferences, in situations of incomplete information. Nearly every investigation involves inferring the unobserved (who committed the crime) from the observed. And here, as in science, some inferences are more rational than others. The investigation could have been botched, or the "evidence" might simply have been fabricated by the police. But there is no way to decide *a priori*, independently of the circumstances, what distinguishes a good investigation from a bad one. Nor can anyone give an absolute guarantee that a particular investigation has yielded the correct result. Moreover, no one can write a definitive treatise on *The Logic of Criminal Investigation*. Nevertheless—and this is the main point—no one doubts that, for some investigations at least (the best ones), the result does indeed correspond to reality. Furthermore, history has permitted us to develop certain rules for conducting an investigation: no one believes anymore in trial by fire, and we doubt the reliability of confessions obtained under torture. It is crucial to compare testimonies, to cross-examine witnesses, to search for physical evidence, etc. Even though there does not exist a methodology based on unquestionable *a priori* reasoning, these rules (and many others) are not arbitrary. They are rational and are based on a detailed analysis of prior experience. In our view, the "scientific method" is not radically different from this kind of approach.

The absence of any "absolutist" criteria of rationality, independent of all circumstances, implies also there is no *general* justification of the principle of induction (another problem going back to Hume). Quite simply, some inductions are justified and others are not; or, to be more precise, some inductions are more reasonable and others are less so. Everything depends on the case at hand: to take a classic philosophical example, the fact that we have seen the Sun rise every day, together with all our astronomical knowledge, gives us good reasons to believe that it will rise tomorrow. But this does not imply that it will rise ten billion years from now (indeed, current astrophysical theories predict that it will exhaust its fuel before then).

In a sense, we always return to Hume's problem: No statement about the real world can ever literally be *proven;* but to use the eminently appropriate expression from Anglo-Saxon law, it can sometimes be proven beyond any *reasonable* doubt. The unreasonable doubt subsists.

If we have spent so much time on these rather elementary remarks, it is because much of the relativist drift that we shall criticize has a double origin:

- Part of twentieth-century epistemology (the Vienna Circle, Popper, and others) has attempted to formalize the scientific method.

- The partial failure of this attempt has led, in some circles, to an attitude of unreasonable skepticism.

In the rest of this chapter, we intend to show that a series of relativist arguments concerning scientific knowledge are either (a) valid critiques of some attempts to formalize the scientific method, which do not, however, in any way undermine the rationality of the scientific enterprise; or (b) mere reformulations, in one guise or another, of Humean radical skepticism.

Epistemology in Crisis

> Science without epistemology is—insofar as it is thinkable at all—
> primitive and muddled. However, no sooner has the
> epistemologist, who is seeking a clear system, fought his way
> through such a system, than he is inclined to interpret the thought-
> content of science in the sense of his system and to reject
> whatever does not fit into his system. The scientist, however,
> cannot afford to carry his striving for epistemological systematic

that far. . . . He therefore must appear to the systematic
epistemologist as an unscrupulous opportunist.

—*Albert Einstein (1949, p. 684)*

Much contemporary skepticism claims to find support in
the writings of philosophers such as Quine, Kuhn, or Feyerabend
who have called into question the epistemology of the first half
of the twentieth century. This epistemology is indeed in crisis.
In order to understand the nature and the origin of the crisis and
the impact that it may have on the philosophy of science, let us
go back to Popper.[62] Of course, Popper is not a relativist, quite
the contrary. He is nevertheless a good starting point, first of all
because many of the modern developments in epistemology
(Kuhn, Feyerabend) arose in reaction to him, and secondly because, while we disagree strongly with some of the conclusions
reached by Popper's critics such as Feyerabend, it is nevertheless true that a significant part of our problems can be traced to
ambiguities or inadequacies in Popper's *The Logic of Scientific
Discovery*.[63] It is important to understand the limitations of this
work in order to face more effectively the irrationalist drift created by the critiques it provoked.

Popper's basic ideas are well known. He wants, first of all,
to give a criterion for demarcating between scientific and non-
scientific theories, and he thinks he has found it in the notion of
falsifiability: in order to be scientific, a theory must make predictions that can, in principle, be false in the real world. For
Popper, theories such as astrology or psychoanalysis avoid sub-
jecting themselves to such a test, either by not making precise
predictions or by arranging their statements in an *ad hoc* fash-

---

[62]We could go back to the Vienna Circle, but that would take us too far afield. Our
analysis in this section is inspired in part by Putnam (1974), Stove (1982), and Laudan
(1990b). After our book appeared in French, Tim Budden drew our attention to
Newton-Smith (1981), where a similar critique of Popper's epistemology can be
found.

[63]Popper (1959).

ion in order to accommodate empirical results whenever they contradict the theory.[64]

If a theory is falsifiable, hence scientific, it may be subjected to attempts at *falsification*. That is, one may compare the theory's empirical predictions with observations or experiments; and if the latter contradict the predictions, it follows that the theory is false and must be rejected. This emphasis on falsification (as opposed to verification) underlines, according to Popper, a crucial asymmetry: one can never prove that a theory is *true*, because it makes, in general, an infinite number of empirical predictions, of which only a finite subset can ever be tested; but one can nevertheless prove that a theory is *false*, because, to do that, a single (reliable) observation contradicting the theory suffices.[65]

The Popperian scheme—falsifiability and falsification—is not a bad one, if it is taken with a grain of salt. But numerous difficulties spring up as soon as one tries to take falsificationist doctrine literally. It may appear attractive to abandon the uncertainty of verification in favor of the certainty of falsification. But this approach runs into two problems: by abandoning verification, one pays too high a price; and one fails to obtain what is promised, because falsification is much less certain than it seems.

The first difficulty concerns the status of scientific induction. When a theory successfully withstands an attempt at falsification, a scientist will, quite naturally, consider the theory to be partially confirmed and will accord it a greater likelihood or a higher subjective probability. The degree of likelihood depends, of course, upon the circumstances: the quality of the experiment, the unexpectedness of the result, etc. But Popper will

---

[64]As we shall see below, whether an explanation is *ad hoc* or not depends strongly upon the context.

[65]In this brief summary we have, of course, grossly oversimplified Popper's epistemology: we have glossed over the distinction between observations, the Vienna-Circle notion of observation statements (which Popper criticizes), and Popper's notion of basic statements; we have omitted Popper's qualification that only *reproducible* effects can lead to falsification; and so forth. However, nothing in the subsequent discussion will be affected by these simplifications.

have none of this: throughout his life, he was a stubborn opponent of any idea of "confirmation" of a theory, or even of its "probability". He wrote:

> *Are we rationally justified in reasoning from repeated instances of which we have experience to instances we have had no experience?* Hume's unrelenting answer is: No, we are not justified . . . My own view is that Hume's answer to this problem is right. (Popper 1974, pp. 1018–1019, italics in the original)[66]

Obviously, every induction is an inference from the observed to the unobserved, and no such inference can be justified using solely *deductive* logic. But, as we have seen, if this argument were to be taken seriously—if rationality were to consist only of deductive logic—it would imply also that there is no good reason to believe that the Sun will rise tomorrow, and yet no one *really* expects the Sun not to rise.

With his method of falsification, Popper thinks that he has solved Hume's problem[67], but his solution, taken literally, is a purely negative one: we can be certain that some theories are false, but never that a theory is true or even probable. Clearly, this "solution" is unsatisfactory from a scientific point of view. In particular, at least one of the roles of science is to make predictions on which other people (engineers, doctors, . . .) can reliably base their activities, and all such predictions rely on some form of induction.

Besides, the history of science teaches us that scientific theories come to be accepted above all because of their successes.

---

[66]See also Stove (1982, p. 48) for similar quotes. Note that Popper calls a theory "corroborated" whenever it successfully passes falsification tests. But the meaning of this word is unclear; it cannot just be a synonym of "confirmed", for otherwise the entire Popperian critique of induction would be empty. See Putnam (1974) for a more detailed discussion.

[67]For example, he writes: "The proposed criterion of demarcation also leads us to a solution of Hume's problem of induction—of the problem of the validity of natural laws. . . . [T]he method of falsification presupposes no inductive inference, but only the tautological transformations of deductive logic whose validity is not in dispute." (Popper 1959, p. 42)

For example, on the basis of Newtonian mechanics, physicists
have been able to deduce a great number of both astronomical
and terrestrial motions, in excellent agreement with observations. Moreover, the credibility of Newtonian mechanics was
reinforced by correct predictions such as the return of Halley's
comet in 1759[68] and by spectacular discoveries such as finding
Neptune in 1846 where Le Verrier and Adams predicted it
should be.[69] It is hard to believe that such a simple theory could
predict so precisely *entirely new* phenomena if it were not at
least approximately true.

The second difficulty with Popper's epistemology is that falsification is much more complicated than it seems.[70] To see this,
let us take once again the example of Newtonian mechanics[71],
understood as the combination of two laws: the law of motion,
according to which force is equal to mass times acceleration;
and the law of universal gravitation, according to which the
force of attraction between two bodies is proportional to the
product of their masses and inversely proportional to the square
of the distance separating them. In what sense is this theory falsifiable? By itself, it doesn't predict much; indeed, a great variety of motions are *compatible* with the laws of Newtonian
mechanics and even *deducible* from them, if one makes appropriate assumptions about the masses of the various celestial
bodies. For example, Newton's famous deduction of Kepler's
laws of planetary motion requires certain *additional assumptions*, which are logically independent of the laws of Newtonian
mechanics, principally that the masses of the planets are small

---

[68]As Laplace wrote: "The learned world awaited with impatience this return which was
to confirm one of the greatest discoveries that have been made in the sciences . . ."
(Laplace 1902 [1825], p. 5)

[69]For a detailed history, see, for example, Grosser (1962) or Moore (1996, chapters 2
and 3).

[70]Let us emphasize that Popper himself is perfectly aware of the ambiguities
associated with falsification. What he does not do, in our opinion, is to provide a
satisfactory alternative to "naive falsificationism"—that is, one which would correct
its defects while retaining at least some of its virtues.

[71]See, for example, Putnam (1974). See also the reply of Popper (1974, pp. 993–999)
and the response of Putnam (1978).

relative to the mass of the Sun: this implies that the mutual interactions between the planets can be neglected, in a first approximation. But this hypothesis, while reasonable, is by no means self-evident: the planets could be made of a very dense material, in which case the additional hypothesis would fail. Or there could exist a large amount of invisible matter affecting the motion of the planets.[72] Moreover, the interpretation of any astronomical observation depends on certain theoretical propositions, in particular on optical hypotheses concerning the functioning of telescopes and the propagation of light through space. The same is true, in fact, for any observation: for example, when one "measures" an electrical current, what one really sees is the position of a needle on a screen (or numbers on a digital readout), which is interpreted, in accordance with our theories, as indicating the presence and the magnitude of a current.[73]

It follows that scientific propositions cannot be falsified one by one, because to deduce from them any empirical proposition whatsoever, it is necessary to make numerous additional assumptions, if only on the way measuring devices work; moreover, these hypotheses are often implicit. The American philosopher Quine has expressed this idea in a rather radical fashion:

[O]ur statements about the external world face the tribunal of sense experience not individually but only as a corporate body. . . . Taken collectively, science has its double dependence upon language and experience; but this duality is not significantly traceable into the statements of science taken one by one. . . .

The idea of defining a symbol in use was . . . an advance over the impossible term-by-term empiricism of Locke and Hume. The statement, rather than the term, came with Ben-

---

[72]Note that the existence of such "dark" matter—invisible, though not necessarily undetectable by other means—is postulated in some contemporary cosmological theories, and these theories are not declared unscientific *ipso facto*.

[73]The importance of theories in the interpretation of experiments has been emphasized by Duhem (1954 [1914], second part, chapter VI).

tham to be recognized as the unit accountable to an empiricist critique. But what I am now urging is that even in taking the statement as unit we have drawn our grid too finely. The unit of empirical significance is the whole of science. (Quine 1980 [1953], pp. 41–42)[74]

What can one reply to such objections? First of all, it must be emphasized that scientists, in their practice, are perfectly aware of the problem. Each time an experiment contradicts a theory, scientists ask themselves a host of questions: Is the error due to the way the experiment was performed or analyzed? Is it due to the theory itself, or to some additional assumption? The experiment itself never dictates what must be done. The notion (what Quine calls the "empiricist dogma") that scientific propositions can be tested one by one belongs to a fairy tale about science.

But Quine's assertions demand serious qualifications.[75] In practice, experience is not given; we do not simply contemplate the world and then interpret it. We perform specific experiments, motivated by our theories, precisely in order to test the different parts of those theories, if possible independently of one another or, at least, in different combinations. We use a *set* of tests, some of which serve only to check that the measuring devices indeed work as expected (by applying them to well-known situations). And, just as it is the totality of the relevant theoretical propositions that is subjected to a falsification test,

---

[74]Let us emphasize that, in the foreword to the 1980 edition, Quine disavows the most radical reading of this passage, saying (correctly in our view) that "empirical content is shared by the statements of science in clusters and cannot for the most part be sorted out among them. Practically the relevant cluster is indeed never the whole of science" (p. viii).

[75]As do some of Quine's related assertions, such as: "Any statement can be held true come what may, if we make drastic enough adjustments elsewhere in the system. Even a statement very close to the periphery [i.e. close to direct experience] can be held true in the face of recalcitrant experience by pleading hallucination or by amending certain statements of the kind called logical laws." (p. 43) Though this passage, taken out of context, might be read as an apologia for radical relativism, Quine's discussion (pp. 43–44) suggests that this is *not* his intention, and that he thinks (again correctly in our view) that certain modifications of our belief systems in the face of "recalcitrant experiences" are much more reasonable than others.

so it is the totality of our empirical observations that constrains our theoretical interpretations. For example, while it is true that our astronomical knowledge depends upon hypotheses about optics, these hypotheses cannot be modified in an arbitrary way, because they can be tested, at least in part, by numerous *independent* experiments.

We have not, however, reached the end of our troubles. If one takes the falsificationist doctrine literally, one should declare that Newtonian mechanics was falsified already in the mid-nineteenth century by the anomalous behavior of Mercury's orbit.[76] For a strict Popperian, the idea of putting aside certain difficulties (such as the orbit of Mercury) in the hope that they will be temporary amounts to an illegitimate strategy aimed at evading falsification. However, if one takes into account the context, one may very well maintain that it is *rational* to proceed in this way, at least for a certain period of time—otherwise science would be impossible. There are always experiments or observations that cannot be fully explained, or that even contradict the theory, which are put aside awaiting better days. Given the immense successes of Newtonian mechanics, it would have been unreasonable to reject it because of a single prediction (apparently) refuted by observations, since this disagreement could have all sorts of other explanations.[77] Science is a rational enterprise, but difficult to codify.

---

[76]Astronomers, beginning with Le Verrier in 1859, noticed that the observed orbit of the planet Mercury differs slightly from the orbit predicted by Newtonian mechanics: the discrepancy corresponds to a precession of the perihelion (point of closest approach to the Sun) of Mercury by approximately 43 seconds of arc per century. (This is an incredibly small angle: recall that one second of arc is 1/3600 of a degree, and one degree is 1/360 of the entire circle.) Various attempts were made to explain this anomalous behavior within the context of Newtonian mechanics: for example, by conjecturing the existence of a new intra-Mercurial planet (a natural idea, given the success of this approach with regard to Neptune). However, all attempts to detect this planet failed. The anomaly was finally explained in 1915 as a consequence of Einstein's general theory of relativity. For a detailed history, see Roseveare (1982).

[77]Indeed, the error could have been in one of the additional hypotheses and not in Newton's theory itself. For example, the anomalous behavior of Mercury's orbit could have been caused by an unknown planet, a ring of asteroids, or a small asphericity of the Sun. Of course, these hypotheses can and should be subjected to tests independent of Mercury's orbit; but these tests depend in turn on additional hypotheses (concerning, for example, the difficulty of seeing a planet close to the

Without a doubt, Popper's epistemology contains some valid insights: the emphasis on falsifiability and falsification is salutary, provided it is not taken to extremes (e.g. the blanket rejection of induction). In particular, when comparing radically different endeavors such as astronomy and astrology, it is useful, to some extent, to employ Popperian criteria. But there is no point in demanding that the pseudo-sciences follow strict rules that the scientists themselves do not follow literally (otherwise one exposes oneself to Feyerabend's critiques, which we shall discuss later).

It is obvious that, in order to be scientific, a theory must be tested empirically in one way or another—and the more stringent the tests, the better. It is also true that predictions of unexpected phenomena often constitute the most spectacular tests. Finally, it is easier to show that a precise quantitative claim is false than to show that it is true. And it is probably a combination of these three ideas that explains, in part, Popper's popularity among many scientists. But these ideas are not due to Popper, nor do they constitute what is original in his work. The necessity of empirical tests goes back at least to the seventeenth century, and is simply the lesson of empiricism: the rejection of *a priori* or revealed truths. Besides, predictions are not always the most powerful tests[78]; and those tests may take relatively complex forms, which cannot be reduced to the simple falsification of hypotheses taken one by one.

Sun) that are not easy to evaluate. We are by no means suggesting that one can continue in this way *ad infinitum*—after a while, the *ad hoc* explanations become too bizarre to be acceptable—but this process may easily take half a century, as it did with Mercury's orbit (see Roseveare 1982).

Besides, Weinberg (1992, pp. 93–94) notes that at the beginning of the twentieth century there were *several* anomalies in the mechanics of the solar system: not only in Mercury's orbit, but also in the orbits of the Moon and of Halley's and Encke's comets. We know now that the latter anomalies were due to errors in the additional hypotheses—the evaporation of gases from comets and the tidal forces acting on the Moon were imperfectly understood—and that only Mercury's orbit constituted a true falsification of Newtonian mechanics. But this was not at all evident at the time.

[78]For example, Weinberg (1992, pp. 90–107) explains why the *retro*diction of the orbit of Mercury was a much more convincing test of general relativity than the *pre*diction of the deflection of starlight by the Sun. See also Brush (1989).

All these problems would not be so serious had they not given rise to a strongly irrationalist reaction: some thinkers, notably Feyerabend, reject Popper's epistemology for many of the reasons just discussed, and then fall into an extreme anti-scientific attitude (see below). But the rational arguments in favor of the theory of relativity or the theory of evolution are to be found in the works of Einstein, Darwin, and their successors, not Popper. Thus, even if Popper's epistemology were entirely false (which is certainly not the case), that would imply nothing concerning the validity of scientific theories.[79]

The Duhem-Quine Thesis: Underdetermination

Another idea, often called the "Duhem-Quine thesis", is that theories are underdetermined by evidence.[80] The set of all our experimental data is finite; but our theories contain, at least potentially, an infinite number of empirical predictions. For example, Newtonian mechanics describes not only how the planets move, but also how a yet-to-be-launched satellite will move. How can one pass from a finite set of data to a potentially infinite set of assertions? Or, to be more precise, is there a unique way of doing this? This is rather like asking whether, given a finite set of points, there is a unique curve that passes through these points. Clearly the answer is no: there are infinitely many curves passing through any given finite set of points. Similarly, there is always a large (even infinite) number of theories compatible with the data—and this, whatever the data and whatever their number.

---

[79]By way of analogy, consider Zeno's paradox: it does not show that Achilles will not, in actual fact, catch the tortoise; it shows only that the concepts of motion and limit were not well understood in Zeno's time. Likewise, we may very well practice science without necessarily understanding how we do it.

[80]Let us emphasize that Duhem's version of this thesis is much less radical than that of Quine. Note also that the term "Duhem-Quine thesis" is sometimes used to designate the idea (analyzed in the previous section) that observations are theory-laden. See Laudan (1990b) for a more detailed discussion of the ideas in this section.

There are two ways to react to such a general thesis. The first approach is to apply it systematically to *all* our beliefs (as one is logically entitled to do). So we would conclude, for example, that, whatever the facts, there will always be just as many suspects at the end of any criminal investigation as there were at the beginning. Clearly, this looks absurd. But it is indeed what can be "shown" using the underdetermination thesis: one can always invent a story (possibly a very bizarre one) in which X is guilty and Y is innocent and in which "the data are accounted for" in an *ad hoc* fashion. We are simply back to Humean radical skepticism. The weakness of this thesis is again its generality.

Another way to deal with this problem is to consider the various concrete situations that can occur when one confronts theory with evidence:

1) One may possess evidence in favor of a given theory that is so strong that to doubt the theory would be almost as unreasonable as to believe in solipsism. For example, we have good reasons to believe that blood circulates, that biological species have evolved, that matter is composed of atoms, and a host of other things. The analogous situation, in a criminal investigation, is that in which one is sure, or almost sure, of having found the culprit.

2) One may have a number of competing theories, none of which seems totally convincing. For example, the problem of the origin of life provides (at least at present) a good example of such a situation. The analogy in criminal investigations is obviously the case in which there are several plausible suspects but it is unclear which one is really guilty. The situation may also arise in which one has just one theory, which, however, is not very convincing due to the lack of sufficiently powerful tests. In such a case, scientists implicitly apply the underdetermination thesis: since another theory, not yet

conceived, might well be the right one, one confers on the sole existing theory a rather low subjective probability.

3) Finally, one may lack even a single plausible theory that accounts for all the existing data. This is probably the case today for the unification of general relativity with elementary-particle physics, as well as for many other difficult scientific problems.

Let us come back for a moment to the problem of the curve drawn through a finite number of points. What convinces us most strongly that we found the right curve is, of course, that when we perform additional experiments, the *new* data fit the *old* curve. One has to assume implicitly that there is not a cosmic conspiracy in which the real curve is very different from the curve we have drawn, but in which all our data (old and new) happen to fall on the intersection of the two. To take a phrase from Einstein, one must imagine that the Lord is subtle, but not malicious.

Kuhn and the Incommensurability of Paradigms

> Much more is known now than was known fifty years ago, and much more was known then than in 1580. So there has been a great accumulation or growth of knowledge in the last four hundred years. This is an extremely well-known fact . . . So a writer whose position inclined him to deny [it], or even made him at all reluctant to admit it, would inevitably seem, to the philosophers who read him, to be maintaining something extremely implausible.
> —David Stove, *Popper and After* (1982, p. 3)

Let us now turn our attention towards some historical analyses that have apparently provided grist for the mill of contemporary relativism. The most famous of these is undoubtedly

Thomas Kuhn's *The Structure of Scientific Revolutions.*[81] We shall deal here exclusively with the epistemological aspect of Kuhn's work, putting aside the details of his historical analyses.[82] There is no doubt that Kuhn envisions his work as a historian as having an impact on our conception of scientific activity and thus, at least indirectly, on epistemology.[83]

Kuhn's scheme is well known: The bulk of scientific activity—what Kuhn calls "normal science"—takes place within "paradigms", which define what kinds of problems are studied, what criteria are used to evaluate a solution, and what experimental procedures are deemed acceptable. From time to time, normal science enters into crisis—a "revolutionary" period—and the paradigm changes. For instance, the birth of modern physics with Galileo and Newton constituted a rupture with Aristotle; similarly, in the twentieth century, relativity theory and quantum mechanics have overturned the Newtonian paradigm. Comparable revolutions took place in biology, during the development from a static view of species to the theory of evolution, or from Lamarck to modern genetics.

This vision of things fits so well with scientists' perception of their own work that it is difficult to see, at first glance, what is revolutionary in this approach, much less how it could be used for anti-scientific purposes. The problem arises only when one faces the notion of the *incommensurability* of paradigms. On the one hand, scientists think, in general, that it is possible to decide rationally between competing theories (Newton and Einstein, for example) on the basis of observations and experiments, even if those theories are accorded the status of "para-

---

[81]For this section, see Shimony (1976), Siegel (1987), and especially Maudlin (1996) for more detailed critiques.

[82]We shall also limit ourselves to *The Structure of Scientific Revolutions* (Kuhn 1962, 2nd ed. 1970). For two quite different analyses of Kuhn's later ideas, see Maudlin (1996) and Weinberg (1996b, p. 56).

[83]Speaking of "the image of science by which we are now possessed" and which is propagated, among others, by scientists themselves, he writes: "This essay attempts to show that we have been misled . . . in fundamental ways. Its aim is a sketch of the quite different concept of science that can emerge from the historical record of the research activity itself." (Kuhn 1970, p. 1)

digms".[84] By contrast, though one can give several meanings to the word "incommensurable" and a good deal of the debate about Kuhn's work has centered on this question, there is at least one version of the incommensurability thesis that casts doubt on the possibility of rational comparison between competing theories, namely the idea that our experience of the world is radically conditioned by our theories, which in turn depend upon the paradigm.[85] For example, Kuhn observes that chemists after Dalton reported chemical compositions as ratios of integers rather than as decimals.[86] And while the atomic theory accounted for much of the data available at that time, some experiments gave conflicting results. The conclusion drawn by Kuhn is rather radical:

> Chemists could not, therefore, simply accept Dalton's theory on the evidence, for much of that was still negative. Instead, even after accepting the theory, they still had to beat nature into line, a process which, in the event, took almost another generation. When it was done, even the percentage composition of well-known compounds was different. The data themselves had changed. That is the last of the senses in which we may want to say that after a revolution scientists work in a different world. (Kuhn 1970, p. 135)

But what exactly does Kuhn mean by, "they still had to beat nature into line"? Is he suggesting that chemists after Dalton manipulated their data in order to make them agree with the atomic hypothesis, and that their successors keep on doing so today? And that the atomic hypothesis is false? Obviously, this is not what Kuhn thinks, but at the very least it is fair to say that

---

[84]Of course, Kuhn does not explicitly deny this possibility, but he tends to emphasize the less empirical aspects that enter into the choice between theories: for example, that "sun worship . . . helped make Kepler a Copernician" (Kuhn 1970, p. 152).

[85]Note that this assertion is much more radical than Duhem's idea that observation depends *in part* on additional theoretical hypotheses.

[86]Kuhn (1970, pp. 130–135).

he has expressed himself in an ambiguous way.[87] It is likely that the measurements of chemical compositions available in the nineteenth century were rather imprecise, and it is possible that the experimenters were so strongly influenced by the atomic theory that they considered it better confirmed than it actually was. Nevertheless, we have *today* so much evidence in favor of atomism (much of which is independent of chemistry) that it has become irrational to doubt it.

Of course, historians have the perfect right to say that this is not what interests them: their aim is to understand what happened when the change of paradigm occurred.[88] And it is interesting to see to what extent that change was based on solid empirical arguments or on extra-scientific beliefs such as sun worship. In an extreme case, a correct change of paradigm may even have occurred, by fortunate accident, for completely irrational reasons. This would in no way alter the fact that the theory originally adopted for faulty reasons is *today* empirically established beyond any reasonable doubt. Furthermore, changes of paradigm, at least in most cases since the birth of modern science, have not occurred for *completely* irrational reasons. The writings of Galileo or Harvey, for instance, contain many empirical arguments and they are by no means all wrong. There is always, to be sure, a complex mixture of good and bad reasons that lead to the emergence of a new theory, and scientists' adherence to the new paradigm may very well have taken place before the empirical evidence became totally convincing.

---

[87]Note also that Kuhn's phrasing—"the percentage composition was different"— confuses *facts* with our *knowledge* of them. What changed, of course, is the chemists' knowledge of (or beliefs about) the percentages, not the percentages themselves.

[88]The historian thus rightly rejects "Whig history": the history of the past rewritten as a forward march toward the present. However, this quite reasonable attitude ought not be confused with another, and rather dubious, methodological proscription, namely the refusal to use all the information available today (including scientific evidence) in order to draw the best possible inferences concerning history, on the pretext that this information was unavailable in the past. After all, art historians utilize contemporary physics and chemistry in order to determine provenance and authenticity; and these techniques are useful for art history even if they were unavailable in the era under study. For an example of similar reasoning in the history of science, see Weinberg (1996a, p. 15).

This is not at all surprising: scientists must try to guess, as best they can, which paths to follow—life is, after all, short—and provisional decisions must often be taken in the absence of sufficient empirical evidence. This does not undermine the rationality of the scientific enterprise, but it does contribute to making the history of science so fascinating.

The basic problem is that there are, as the philosopher of science Tim Maudlin has eloquently pointed out, *two* Kuhns—a moderate Kuhn and his immoderate brother—jostling elbows throughout the pages of *The Structure of Scientific Revolutions*. The moderate Kuhn admits that the scientific debates of the past were settled correctly, but emphasizes that the evidence available at the time was weaker than is generally thought and that non-scientific considerations played a role. We have no objection of principle to the moderate Kuhn, and we leave to historians the task of investigating the extent to which these ideas are correct in concrete situations.[89] By contrast, the immoderate Kuhn—who became, perhaps involuntarily, one of the founding fathers of contemporary relativism—thinks that changes of paradigm are due principally to non-empirical factors and that, once accepted, they condition our perception of the world to such an extent that they can *only* be confirmed by our subsequent experiences. Maudlin eloquently refutes this idea:

If presented with a moon rock, Aristotle would experience it as a rock, and as an object with a tendency to fall. He could not fail to conclude that the material of which the moon is made is not fundamentally different from terrestrial material with respect to its natural motion.[90] Similarly, ever better telescopes revealed more clearly the phases of Venus, irrespec-

---

[89]See, for example, the studies in Donovan *et al.* (1988).

[90][This note and the two following are added by us.] According to Aristotle, terrestrial matter is made of four elements—fire, air, water and earth—whose natural tendency is to rise (fire, air) or to fall (water, earth) according to their composition; while the Moon and other celestial bodies are made of a special element, "aether", whose natural tendency is to follow a perpetual circular motion.

tive of one's preferred cosmology[91], and even Ptolemy would
have remarked the apparent rotation of a Foucault pendulum.[92] The sense in which one's paradigm may influence one's
experience of the world cannot be so strong as to guarantee
that one's experience will always accord with one's theories,
else the need to revise theories would never arise. (Maudlin
1996, p. 442[93])

Thus, while it is true that scientific experiments do not provide
their own interpretation, it is also true that the theory does not
determine the perception of the results.

The second objection against the radical version of Kuhn's
history of science—an objection we shall also use later against
the "strong programme" in the sociology of science—is that of

---

[91]Ever since antiquity, it was observed that Venus is never very far from the Sun in the
sky. In Ptolemy's geocentric cosmology, this was explained by supposing *ad hoc* that
Venus and the Sun revolve more or less synchronously around the Earth (Venus being
closer). It follows that Venus should be seen always as a thin crescent, like the "new
moon". On the other hand, the heliocentric theory accounts naturally for the
obervations by supposing that Venus orbits the Sun at a smaller radius than the Earth.
It follows that Venus should, like the Moon, exhibit phases ranging from "new" (when
Venus is on the same side of the Sun as the Earth) to almost "full" (when Venus is on
the far side of the Sun). Since Venus appears to the naked eye as a point, it was not
possible to distinguish empirically between these two predictions until telescopic
observations by Galileo and his successors clearly established the existence of the
phases of Venus. While this did not *prove* the heliocentric model (other theories were
also able to explain the phases), it did give significant evidence in its favor, as well as
strong evidence against the Ptolemaic model.

[92]According to Newtonian mechanics, a swinging pendulum remains always in a
single plane; this prediction holds, however, only with respect to a so-called "inertial
frame of reference", such as one fixed with respect to the distant stars. A frame of
reference attached to the Earth is *not* precisely inertial, due to the Earth's daily
rotation around its axis. The French physicist Jean Bernard Léon Foucault
(1819–1868) realized that the direction of swinging of a pendulum, seen relative to the
Earth, would gradually precess, and that this can be understood as evidence for the
Earth's rotation. To see this, consider, for example, a pendulum located at the north
pole. Its direction of swing will remain fixed relative to the distant stars, while the
Earth rotates underneath it; therefore, *relative to an observer on the Earth*, its
direction of swing will make one full rotation every 24 hours. At all other latitudes
(except the equator), a similar effect holds but the precession is slower: for example,
at the latitude of Paris (49° N), the precession is once every 32 hours. In 1851
Foucault demonstrated this effect, using a pendulum 67 meters long hung from the
dome of the Panthéon. Shortly thereafter, the Foucault pendulum became a standard
demonstration in science museums around the world.

[93]This essay has thus far been published only in French translation. We thank
Professor Maudlin for supplying us with the original English text.

self-refutation. Research in history, and in particular in the history of science, employs methods that are not radically different from those used in the natural sciences: studying documents, drawing the most rational inferences, making inductions based on the available data, and so forth. If arguments of this type in physics or biology did not allow us to arrive at reasonably reliable conclusions, what reason would there be to trust them in history? Why speak in a realist mode about historical categories, such as paradigms, if it is an illusion to speak in a realist mode about scientific concepts (which are in fact much more precisely defined) such as electrons or DNA?[94]

But one may go further. It is natural to introduce a hierarchy in the degree of credence accorded to different theories, depending on the quantity and quality of the evidence supporting them.[95] Every scientist—indeed, every human being—proceeds in this way and grants a higher subjective probability to the best-established theories (for instance, the evolution of species or the existence of atoms) and a lower subjective probability to more speculative theories (such as detailed theories of quantum gravity). The same reasoning applies when comparing theories in natural science with those in history or sociology. For example, the evidence of the Earth's rotation is vastly stronger than anything Kuhn could put forward in support of his historical theories. This does not mean, of course, that physicists are more clever than historians or that they use better methods, but simply that they deal with less complex problems, involving a smaller number of variables which, moreover, are easier to measure and to control. It is impossible to avoid introducing such a hierarchy in our beliefs, and this hierarchy implies that there is no conceivable argument based on the Kuhnian view of history

---

[94]It is worth noting that a similar argument was put forward by Feyerabend in the last edition of *Against Method*: "It is not enough to undermine the authority of the sciences by historical arguments: why should the authority of history be greater than that of, say, physics?" (Feyerabend 1993, p. 271) See also Ghins (1992, p. 255) for a similar argument.

[95]This type of reasoning goes back at least to Hume's argument against miracles: see Hume (1988 [1748], section X).

that could give succor to those sociologists or philosophers who wish to challenge, in a blanket way, the reliability of scientific results.

Feyerabend: "Anything Goes"

Another famous philosopher who is often quoted in discussions about relativism is Paul Feyerabend. Let us begin by acknowledging that Feyerabend is a complicated character. His personal and political attitudes have earned him a fair amount of sympathy, and his criticisms of attempts at codifying scientific practice are often justified. Moreover, despite the title of one of his books, *Farewell to Reason*, he never became entirely and openly irrationalist; towards the end of his life he started to distance himself (or so it seems) from the relativist and anti-scientific attitudes of some of his followers.[96] Nevertheless, Feyerabend's writings contain numerous ambiguous or confused statements, which sometimes end in violent attacks against modern science: attacks which are simultaneously philosophical, historical, and political, and in which judgments of fact are mixed with judgments of value.[97]

The main problem in reading Feyerabend is to know when to take him seriously. On the one hand, he is often considered as a sort of court jester in the philosophy of science, and he

---

[96]For example, in 1992 he wrote:

> How can an enterprise [science] depend on culture in so many ways, and yet produce such solid results? . . . Most answers to this question are either incomplete or incoherent. Physicists take the fact for granted. Movements that view quantum mechanics as a turning-point in thought—and that include fly-by-night mystics, prophets of a New Age, and relativists of all sorts—get aroused by the cultural component and forget predictions and technology. (Feyerabend 1992, p. 29)

See also Feyerabend (1993, p. 13, note 12).

[97]See, for example, Chapter 18 of *Against Method* (Feyerabend 1975). This chapter is not, however, included in the later editions of the book in English (Feyerabend 1988, 1993). See also Chapter 9 of *Farewell to Reason* (Feyerabend 1987).

seems to have taken some pleasure in playing this role.[98] At times he himself emphasized that his words ought not be taken literally.[99] On the other hand, his writings are full of references to specialized works in the history and philosophy of science, as well as in physics; and this aspect of his work has greatly contributed to his reputation as a major philosopher of science. Bearing all this in mind, we shall discuss what seem to us to be his fundamental errors, and illustrate the excesses to which they can lead.

We fundamentally agree with what Feyerabend says about the scientific method, considered in the abstract:

> The idea that science can, and should, be run according to fixed and universal rules, is both unrealistic and pernicious. (Feyerabend 1975, p. 295)

He criticizes at length the "fixed and universal rules" through which earlier philosophers thought that they could express the essence of the scientific method. As we have said, it is extremely difficult, if not impossible, to codify the scientific method, though this does not prevent the development of certain rules, with a more-or-less general degree of validity, on the basis of previous experience. If Feyerabend had limited himself to showing, through historical examples, the limitations of any general and universal codification of the scientific method,

---

[98]For example, he writes: "Imre Lakatos, somewhat jokingly, called me an anarchist and I had no objection to putting on the anarchist's mask." (Feyerabend 1993, p. vii)

[99]For example: "the main ideas of [this] essay . . . are rather trivial and appear trivial when expressed in suitable terms. I prefer more paradoxical formulations, however, for nothing dulls the mind as thoroughly as hearing familiar words and slogans." (Feyerabend 1993, p. xiv) And also: "Always remember that the demonstrations and the rhetorics used do not express any 'deep convictions' of mine. They merely show how easy it is to lead people by the nose in a rational way. An anarchist is like an undercover agent who plays the game of Reason in order to undercut the authority of Reason (Truth, Honesty, Justice, and so on)." (Feyerabend 1993, p. 23) This passage is followed by a footnote referring to the Dadaist movement.

we could only agree with him.[100] Unfortunately, he goes much farther:

> All methodologies have their limitations and the only 'rule' that survives is 'anything goes'. (Feyerabend 1975, p. 296)

This is an erroneous inference that is typical of relativist reasoning. Starting from a correct observation—"all methodologies have their limitations"—Feyerabend jumps to a totally false conclusion: "anything goes". There are several ways to swim, and all of them have their limitations, but it is not true that all bodily movements are equally good (if one prefers not to sink). There is no unique method of criminal investigation, but this does not mean that all methods are equally reliable (think about trial by fire). The same is true of scientific methods.

In the second edition of his book, Feyerabend tries to defend himself against a literal reading of "anything goes". He writes:

> A naive anarchist says (a) that both absolute rules and context-dependent rules have their limits and infers (b) that all rules and standards are worthless and should be given up. Most reviewers regard me as a naive anarchist in this sense . . . [But] while I agree with (a) I do not agree with (b). I argue that all rules have their limits and that there is no comprehensive 'rationality', I do not argue that we should proceed without rules and standards. (Feyerabend 1993, p. 231)

---

[100]However, we take no position on the validity of the details of his historical analyses. See, for example, Clavelin (1994) for a critique of Feyerabend's theses concerning Galileo.

Let us note also that several of his discussions of problems in modern physics are erroneous or grossly exaggerated: see, for example his claims concerning Brownian motion (Feyerabend 1993, pp. 27–29), renormalization (p. 46), the orbit of Mercury (pp. 47–49), and scattering in quantum mechanics (pp. 49–50n). To disentangle all these confusions would take too much space; but see Bricmont (1995a, p. 184) for a brief analysis of Feyerabend's claims concerning Brownian motion and the second law of thermodynamics.

The problem is that Feyerabend gives little indication of the *content* of these "rules and standards"; and unless they are constrained by some notion of rationality, one arrives easily at the most extreme form of relativism.

When Feyerabend addresses concrete issues, he frequently mixes reasonable observations with rather bizarre suggestions:

> [T]he first step in our criticism of customary concepts and customary reactions is to step outside the circle and either to invent a new conceptual system, for example a new theory, that clashes with the most carefully established observational results and confounds the most plausible theoretical principles, or to import such a system from outside science, from religion, from mythology, from the ideas of incompetents, or the ramblings of madmen. (Feyerabend 1993, pp. 52–53)[101]

One could defend these assertions by invoking the classical distinction between the context of *discovery* and the context of *justification*. Indeed, in the idiosyncratic process of inventing scientific theories, all methods are in principle admissible— deduction, induction, analogy, intuition and even hallucination[102]—and the only real criterion is pragmatic. On the other hand, the justification of theories must be rational, even if this rationality cannot be definitively codified. One might be tempted to think that Feyerabend's admittedly extreme examples concern solely the context of discovery, and that there is thus no real contradiction between his viewpoint and ours.

But the problem is that Feyerabend explicitly *denies* the validity of the distinction between discovery and justification.[103] Of course, the sharpness of this distinction was greatly exaggerated in traditional epistemology. We always come back to the same problem: it is naive to believe that there exist general,

---

[101]For a similar statement, see Feyerabend (1993, p. 33).

[102]For example, it is said that the chemist Kekule (1829–1896) was led to conjecture (correctly) the structure of benzene as the result of a dream.

[103]Feyerabend (1993, pp. 147–149).

context-independent rules that allow us to verify or falsify a theory; otherwise put, the context of justification and the context of discovery evolve historically in parallel.[104] Nevertheless, at each moment of history, such a distinction exists. If it didn't, the justification of theories would be unconstrained by any considerations of rationality. Let us think again about criminal investigations: the culprit can be discovered thanks to all sorts of fortuitous events, but the evidence put forward to prove his guilt does not enjoy such a freedom (even if the standards of evidence also evolve historically).[105]

Once Feyerabend has made the leap to "anything goes", it is not surprising that he constantly compares science with mythology or religion, as, for example, in the following passage:

> Newton reigned for more than 150 years, Einstein briefly introduced a more liberal point of view only to be succeeded by the Copenhagen Interpretation. The similarities between science and myth are indeed astonishing. (Feyerabend 1975, p. 298)

Here Feyerabend is suggesting that the so-called Copenhagen interpretation of quantum mechanics, due principally to Niels Bohr and Werner Heisenberg, was accepted by physicists in a rather dogmatic way, which is not entirely false. (It is less clear which point of view of Einstein he is alluding to.) But what Feyerabend does not give are examples of myths that change because experiments contradict them, or that suggest experiments aimed at discriminating between earlier and later versions of the myth. It is only for this reason—which is crucial—that the "similarities between science and myth" are superficial.

---

[104]For example, the anomalous behavior of Mercury's orbit acquired a different epistemological status with the advent of general relativity (see notes 76–78 above).

[105]A similar remark can be made about the classical distinction, also criticized by Feyerabend, between observational and theoretical statements. One should not be naive when saying that one "measures" something; nevertheless, there do exist "facts"—for example, the position of a needle on a screen or the characters on a computer printout—and these facts do not always coincide with our desires.

This analogy occurs again when Feyerabend suggests separating Science and the State:

> While the parents of a six-year-old child can decide to have him instructed in the rudiments of Protestantism, or in the rudiments of the Jewish faith, or to omit religious instruction altogether, they do not have a similar freedom in the case of the sciences. Physics, astronomy, history *must* be learned. They cannot be replaced by magic, astrology, or by a study of legends.
>
> Nor is one content with a merely *historical* presentation of physical (astronomical, historical, etc.) facts and principles. One does not say: *some people believe* that the earth moves round the sun while others regard the earth as a hollow sphere that contains the sun, the planets, the fixed stars. One says: the earth *moves* round the sun—everything else is sheer idiocy. (Feyerabend 1975, p. 301)

In this passage Feyerabend reintroduces, in a particularly brutal form, the classical distinction between "facts" and "theories"—a basic tenet of the Vienna Circle epistemology he rejects. At the same time he appears to use implicitly in the social sciences a naively realist epistemology that he rejects for the natural sciences. How, after all, does one find out exactly what "some people believe", if not by using methods analogous to those of the sciences (observations, polls, etc.)? If, in a survey of Americans' astronomical beliefs, the sample were limited to physics professors, there would probably be no one who "regards the earth as a hollow sphere"; but Feyerabend could respond, quite rightly, that the poll was poorly designed and the sampling biased (would he dare say that it is unscientific?). The same goes for an anthropologist who stays in New York and invents in his office the myths of other peoples. But which criteria acceptable to Feyerabend would be violated? Doesn't anything go? Feyerabend's methodological relativism, if taken

literally, is so radical that it becomes self-refuting. Without a
minimum of (rational) method, even a "merely historical presentation of facts" becomes impossible.

What is striking in Feyerabend's writings is, paradoxically,
their abstractness and generality. His arguments show, at best,
that science does not progress by following a well-defined
method, and with that we basically agree. But Feyerabend never
explains in what sense atomic theory or evolution theory might
be *false*, despite all that we know today. And if he does not say
that, it is probably because he does not believe it, and shares (at
least in part) with most of his colleagues the scientific view of
the world, namely that species evolved, that matter is made of
atoms, etc. And if he shares those ideas, it is probably because
he has good reasons to do so. Why not think about those reasons and try to make them explicit, rather than just repeating
over and over again that they are not justifiable by some universal rules of method? Working case by case, he could show
that there are indeed solid empirical arguments supporting
those theories.

Of course, this may or may not be the kind of question that
interests Feyerabend. He often gives the impression that his opposition to science is not of a cognitive nature but follows rather
from a choice of lifestyle, as when he says: "love becomes impossible for people who insist on 'objectivity', i.e. who live entirely in accordance with the spirit of science."[106] The trouble is
that he fails to make a clear distinction between factual judgments and value judgments. He could, for example, maintain
that evolution theory is infinitely more plausible than any creationist myth, but that parents nevertheless have a right to demand that schools teach false theories to their children. We
would disagree, but the debate would no longer be purely on the
cognitive level, and would involve political and ethical considerations.

---

[106]Feyerabend (1987, p. 263).

In the same vein, Feyerabend writes in the introduction to the Chinese edition of *Against Method*[107]:

> *First-world science is one science among many* . . . My main motive in writing the book was humanitarian, not intellectual. I wanted to support people, not to 'advance knowledge'. (Feyerabend 1988, p. 3 and 1993, p. 3, italics in the original)

The problem is that the first thesis is of a purely cognitive nature (at least if he is speaking of science and not of technology), while the second is linked to practical goals. But if, in reality, there are no "other sciences" really distinct from those of the "first world" that are nevertheless equally powerful at the cognitive level, in what way would asserting the first thesis (which would be false) allow him to "support people"? The problems of truth and objectivity cannot be evaded so easily.

The "Strong Programme" in the Sociology of Science

During the 1970s, a new school in the sociology of science arose. While previous sociologists of science were, in general, content to analyze the social context in which scientific activity takes place, the researchers gathered under the banner of the "strong programme" were, as the name indicates, considerably more ambitious. Their aim was to explain in sociological terms the *content* of scientific theories.

Of course, most scientists, when they hear about these ideas, protest and point out the substantial missing piece in this kind of explanation: Nature itself.[108] In this section we shall ex-

---

[107]Reproduced in the second and third English editions.

[108]For case studies in which scientists and historians of science explain the concrete mistakes contained in analyses by supporters of the strong programme, see, for example, Gingras and Schweber (1986), Franklin (1990, 1994), Mermin (1996a, 1996b, 1996c, 1997a), Gottfried and Wilson (1997), and Koertge (1998).

plain the fundamental conceptual problems faced by the strong programme. While some of its supporters have recently made corrections to their initial claims, they do not seem to realize the extent to which their starting point was erroneous.

Let us start by quoting the principles set forth for the sociology of knowledge by one of the founders of the strong programme, David Bloor:

1. It would be causal, that is, concerned with the conditions which bring about belief or states of knowledge. Naturally there will be other types of causes apart from social ones which will cooperate in bringing about belief.

2. It would be impartial with respect to truth and falsity, rationality or irrationality, success or failure. Both sides of these dichotomies will require explanation.

3. It would be symmetrical in its style of explanation. The same types of cause would explain, say, true and false beliefs.

4. It would be reflexive. In principle its patterns of explanation would have to be applicable to sociology itself. (Bloor 1991, p. 7)

To grasp what is meant by "causal", "impartial" and "symmetrical", we shall analyze an article of Bloor and his colleague Barry Barnes in which they explain and defend their programme.[109] The article begins with an apparent statement of good will:

Far from being a threat to the scientific understanding of forms of knowledge, relativism is required by it. . . . It is those who oppose relativism, and who grant certain forms of knowledge a privileged status, who pose the real threat to a scien-

---

[109]Barnes and Bloor (1981).

tific understanding of knowledge and cognition. (Barnes and
Bloor 1981, pp. 21–22)

However, this already raises the issue of self-refutation: Doesn't
the discourse of the sociologist, who wants to provide "a scientific understanding of knowledge and cognition", claim "a privileged status" with respect to other discourses, for example
those of the "rationalists" that Barnes and Bloor criticize in the
rest of their article? It seems to us that, if one seeks to have a
"scientific" understanding of anything, one is forced to make a
distinction between a good and a bad understanding. Barnes
and Bloor seem to be aware of this, since they write:

> The relativist, like everyone else, is under the necessity to
> sort out beliefs, accepting some and rejecting others. He will
> naturally have preferences and these will typically coincide
> with those of others in his locality. The words 'true' and 'false'
> provide the idiom in which those evaluations are expressed,
> and the words 'rational' and 'irrational' will have a similar
> function. (Barnes and Bloor 1981, p. 27)

But this is a strange notion of "truth", which manifestly contradicts the notion used in everyday life.[110] If I regard the statement "I drank coffee this morning" as true, I do not mean simply that I *prefer* to believe that I drank coffee this morning,
much less that "others in my locality" think that I drank coffee
this morning![111] What we have here is a radical redefinition of
the concept of truth, which nobody (starting with Barnes and
Bloor themselves) would accept in practice for ordinary knowledge. Why, then, should it be accepted for scientific knowledge?
Note also that, even in the latter context, this definition doesn't

---

[110]One could of course interpret these words as a mere *description*: people tend to
call "true" what they believe. But, with that interpretation, the statement would be
banal.

[111]This example is adapted from Bertrand Russell's critique of the pragmatism of
William James and John Dewey: see Chapters 24 and 25 of Russell (1961a), in
particular p. 779.

hold water: Galileo, Darwin, and Einstein did not sort out their beliefs by following those of others in their locality.

Moreover, Barnes and Bloor fail to use systematically their new notion of "truth"; from time to time they fall back, without comment, on the traditional sense of the word. For example, at the beginning of their article, they admit that "to say that all beliefs are equally true encounters the problem of how to handle beliefs which contradict one another", and that "to say that all beliefs are equally false poses the problem of the status of the relativist's own claims."[112] But if "a true belief" meant only "a belief that one shares with other people in one's locality", the problem of the contradiction between beliefs held in different places would no longer pose any problem.[113]

A similar ambiguity plagues their discussion of rationality:

> For the relativist there is no sense attached to the idea that some standards or beliefs are really rational as distinct from merely locally accepted as such. (Barnes and Bloor 1981, p. 27)

Again, what exactly does this mean? Isn't it "really rational" to believe that the Earth is (approximately) round, at least for

---

[112]Barnes and Bloor (1981, p. 22).

[113]A similar slippage arises in their use of the word "knowledge". Philosophers usually understand "knowledge" to mean "justified true belief" or some similar concept, but Bloor begins by offering a radical redefinition of the term:

> Instead of defining it as true belief—or perhaps, justified true belief— knowledge for the sociologist is whatever people take to be knowledge. It consists of those beliefs which people confidently hold to and live by. . . . Of course knowledge must be distinguished from mere belief. This can be done by reserving the word 'knowledge' for what is collectively endorsed, leaving the individual and idiosyncratic to count as mere belief. (Bloor 1991, p. 5; see also Barnes and Bloor 1981, p. 22n)

However, only nine pages after enunciating this non-standard definition of "knowledge", Bloor reverts without comment to the standard definition of "knowledge", which he contrasts with "error": "[I]t would be wrong to assume that the natural working of our animal resources always produces knowledge. They produce a mixture of knowledge and error with equal naturalness . . ." (Bloor 1991, p. 14).

those of us who have access to airplanes and satellite photos? Is this merely a "locally accepted" belief?

Barnes and Bloor seem here to be playing on two levels: a general skepticism, which of course cannot be refuted; and a concrete program aiming at a "scientific" sociology of knowledge. But the latter presupposes that one has given up radical skepticism and that one is trying, as best one can, to understand some part of reality.

Let us therefore temporarily put aside the arguments in favor of radical skepticism, and ask whether the "strong programme", considered as a scientific project, is plausible. Here is how Barnes and Bloor explain the symmetry principle on which the strong programme is based:

> Our equivalence postulate is that all beliefs are on a par with one another with respect to the causes of their credibility. It is not that all beliefs are equally true or equally false, but that regardless of truth and falsity the fact of their credibility is to be seen as equally problematic. The position we shall defend is that the incidence of all beliefs without exception calls for empirical investigation and must be accounted for by finding the specific, local causes of this credibility. This means that regardless of whether the sociologist evaluates a belief as true or rational, or as false and irrational, he must search for the causes of its credibility. . . . All these questions can, and should, be answered without regard to the status of the belief as it is judged and evaluated by the sociologist's own standards. (Barnes and Bloor 1981, p. 23)

Here, instead of a *general* skepticism or philosophical relativism, Barnes and Bloor are clearly proposing a *methodological* relativism for sociologists of knowledge. But the ambiguity remains: what exactly do they mean by "without regard to the status of the belief as it is judged and evaluated by the sociologist's own standards"?

If the claim were merely that we should use the same principles of sociology and psychology to explain the causation of all beliefs irrespective of whether we evaluate them as true or false, rational or irrational, then we would have no particular objection.[114] But if the claim is that only *social* causes can enter into such an explanation—that the way the world *is* (i.e., Nature) cannot enter—then we cannot disagree more strenuously.[115]

In order to understand the role of Nature, let us consider a concrete example: Why did the European scientific community become convinced of the truth of Newtonian mechanics sometime between 1700 and 1750? Undoubtedly a variety of historical, sociological, ideological, and political factors must play a part in this explanation—one must explain, for example, why Newtonian mechanics was accepted quickly in England but more slowly in France[116]—but certainly *some* part of the explanation (and a rather important part at that) must be that the planets and comets really do move (to a very high degree of approximation, though not exactly) as predicted by Newtonian mechanics.[117]

---

[114]Though one might have qualms about the hyper-scientistic attitude that human beliefs can always be explained causally, and about the assumption that we have at present adequate and well-verified principles of sociology and psychology that can be used for this purpose.

[115]Elsewhere, Bloor does state explicitly that "Naturally there will be other types of causes apart from social ones which will cooperate in bringing about belief" (Bloor 1991, p. 7). The trouble is that he fails to make explicit *in what way* natural causes will be allowed to enter into the explanation of belief, or what precisely will be left of the symmetry principle if natural causes are taken seriously. For a more detailed critique of Bloor's ambiguities (from a philosophical point of view slightly different from ours), see Laudan (1981); see also Slezak (1994).

[116]See, for example, Brunet (1931) and Dobbs and Jacob (1995).

[117]Or more precisely: There is a vast body of extremely convincing astronomical evidence in support of the belief that the planets and comets do move (to a very high degree of approximation, though not exactly) as predicted by Newtonian mechanics; and *if* this belief is correct, then it is the fact of this motion (and not merely our belief in it) that forms part of the explanation of why the eighteenth-century European scientific community came to believe in the truth of Newtonian mechanics. Please note that *all* our assertions of fact—including "today in New York it's raining"—should be glossed in this way.

Here's a more homely example: Suppose we encounter a man running out of a lecture hall screaming at the top of his lungs that there's a stampeding herd of elephants inside. What we are to make of this assertion, and in particular how we are to evaluate its "causes", should, it seems clear, depend heavily on whether or not there *is* in fact a stampeding herd of elephants in the room—or, more precisely, since admittedly we have no direct, unmediated access to external reality—whether when we and other people peek (cautiously!) into the room *we* see or hear a stampeding herd of elephants (or the destruction that such a herd might recently have caused before exiting the room). If we do see such evidence of elephants, then the most plausible explanation for the entire set of observations is that there *is* (or was) in fact a stampeding herd of elephants in the lecture hall, that the man saw and/or heard it, and that his subsequent fright (which we might well share under the circumstances) led him to exit the room in a hurry and to scream the assertion we overheard. And our reaction would be to call the police and the zookeepers. If, on the other hand, our own observations reveal no evidence of elephants in the lecture hall, then the most plausible explanation is that there was *not* in fact a stampeding herd of elephants in the room, that the man *imagined* the elephants as a result of some psychosis (whether internally or chemically induced), and that *this* led him to exit the room in a hurry and to scream the assertion we overheard. And we'd call the police and the psychiatrists.[118] And we daresay that Barnes and Bloor, whatever they might write in journal articles for sociologists and philosophers, would do the same in real life.

Now, as we explained before, we do not see any *fundamental* difference between the epistemology of science and the

---

[118]For what it's worth, these decisions can presumably be justified on Bayesian grounds, using our prior experience of the probability of finding elephants in lecture halls, of the incidence of psychosis, of the reliability of our own visual and auditory perceptions, and so forth.

rational attitude in everyday life: the former is nothing but the extension and refinement of the latter. Any philosophy of science—or methodology for sociologists—that is so blatantly wrong when applied to the epistemology of everyday life must be severely flawed at its core.

In summary, it seems to us that the "strong programme" is ambiguous in its intent; and, depending on how one resolves the ambiguity, it becomes either a valid and mildly interesting corrective to the most naive psychological and sociological notions—reminding us that "true beliefs have causes too"—or else a gross and blatant error.

The supporters of the "strong programme" thus face a dilemma. They could, if they choose, adhere systematically to a philosophical skepticism or relativism; but in that case it is unclear why (or how) they would seek to build a "scientific" sociology. Alternatively, they could choose to adopt only a methodological relativism; but this position is untenable if one abandons philosophical relativism, because it ignores an essential element of the desired explanation, namely Nature itself. For this reason, the sociological approach of the "strong programme" and the relativistic philosophical attitude are mutually reinforcing. Therein resides the danger (and no doubt the appeal for some) of the different variants of this programme.

Bruno Latour and His Rules of Method

The strong programme in the sociology of science has found an echo in France, particularly around Bruno Latour. His works contain a great number of propositions formulated so ambiguously that they can hardly be taken literally. And when one removes the ambiguity—as we shall do here in a few examples—one reaches the conclusion that the assertion is either true but banal, or else surprising but manifestly false.

In his theoretical work, *Science in Action*[119], Latour develops seven Rules of Method for the sociologist of science. Here is the Third Rule of Method:

> Since the settlement of a controversy is the *cause* of Nature's representation, not the consequence, we can never use the outcome—Nature—to explain how and why a controversy has been settled. (Latour 1987, pp. 99, 258)

Note how Latour slips, without comment or argument, from "Nature's representation" in the first half of this sentence to "Nature" *tout court* in the second half. If we were to read "Nature's representation" in *both* halves, then we'd have the truism that scientists' *representations* of Nature (that is, their theories) are arrived at by a social process, and that the course and outcome of that social process can't be explained simply by its outcome. If, on the other hand, we take "Nature" seriously in the second half, linked as it is to the word "outcome", then we would have the claim that the external world is *created* by scientists' negotiations: a claim that is, to say the least, a rather bizarre form of radical idealism. Finally, if we take "Nature" seriously in the second half but expunge the word "outcome" preceding it, then we would have either (a) the weak (and trivially true) claim that the course and outcome of a scientific controversy cannot be explained *solely* by the nature of the external world (obviously *some* social factors play a role, if only in determining which experiments are technologically feasible at a given time, not to mention other, more subtle social influences); or (b) the strong (and manifestly false) claim that the nature of the external world plays *no* role in constraining the course and outcome of a scientific controversy.[120]

---

[119]Latour (1987). For a more detailed analysis of *Science in Action*, see Amsterdamska (1990). For a critical analysis of the later theses of Latour's school (as well as of other trends in sociology of science), see Gingras (1995).

[120]Re (b), the "homely example" in Gross and Levitt (1994, pp. 57–58) makes the point clearly.

We could be accused here of focusing our attention on an ambiguity of formulation and of not trying to understand what Latour really means. In order to counter this objection, let us go back to the section "Appealing (to) Nature" (pp. 94–100) where the Third Rule is introduced and developed. Latour begins by ridiculing the appeal to Nature as a way of resolving scientific controversies, such as the one concerning solar neutrinos[121]:

> A fierce controversy divides the astrophysicists who calcu-
> late the number of neutrinos coming out of the sun and Davis,
> the experimentalist who obtains a much smaller figure. It is
> easy to distinguish them and put the controversy to rest. Just
> let us see for ourselves in which camp the sun is really to be
> found. Somewhere the natural sun with its true number of
> neutrinos will close the mouths of dissenters and force them
> to accept the facts no matter how well written these papers
> were. (Latour 1987, p. 95)

Why does Latour choose to be ironic? The problem is to know how many neutrinos are emitted by the Sun, and this question is indeed difficult. We can hope that it will be resolved some day, not because "the natural sun will close the mouths of dis-senters", but because sufficiently powerful empirical data will become available. Indeed, in order to fill in the gaps in the currently available data and to discriminate between the currently existing theories, several groups of physicists have recently built detectors of different types, and they are now performing

---

[121]The nuclear reactions that power the Sun are expected to emit copious quantities of the subatomic particle called the neutrino. By combining current theories of solar structure, nuclear physics, and elementary-particle physics, it is possible to obtain quantitative predictions for the flux and energy distribution of the solar neutrinos. Since the late 1960s, experimental physicists, beginning with the pioneering work of Raymond Davis, have been attempting to detect the solar neutrinos and measure their flux. The solar neutrinos have in fact been detected; but their flux appears to be less than one-third of the theoretical prediction. Astrophysicists and elementary-particle physicists are actively trying to determine whether the discrepancy arises from experimental error or theoretical error, and if the latter, whether the failure is in the solar models or in the elementary-particle models. For an introductory overview, see Bahcall (1990).

the (difficult) measurements.[122] It is thus reasonable to expect that the controversy will be settled sometime in the next few years, thanks to an accumulation of evidence that, taken together, will indicate clearly the correct solution. However, other scenarios are in principle possible: the controversy could die out because people stop being interested in the issue, or because the problem turns out to be too difficult to solve; and, at this level, sociological factors undoubtedly play a role (if only because of the budgetary constraints on research). Obviously, scientists think, or at least hope, that if the controversy is resolved it will be because of observations and not because of the literary qualities of the scientific papers. Otherwise, they will simply have ceased to do science.

But we, like Latour, do not work professionally on the solar-neutrino problem; we are unable to render an informed guess as to how many neutrinos the Sun emits. We could try to get a rough idea by examining the scientific literature on the subject; or failing that, we could get an even rougher idea by examining the sociological aspects of the problem, for example, the scientific respectability of the researchers involved in the controversy. And there is no doubt that, in practice, this is what scientists themselves do when they don't work in the field, for lack of a better alternative. But the degree of certainty provided by this kind of investigation is very weak. Nevertheless, Latour seems to accord it a crucial role. He distinguishes between two "versions": according to the first, it is Nature that decides the outcome of controversies; according to the second, the power struggles between researchers play that role.

It is crucial for us, laypeople who want to understand technoscience, to decide which version is right, because in the first version, as Nature is enough to settle all disputes, we have nothing to do since no matter how large the resources of

---

[122]See, for example, Bahcall *et al.* (1996).

the scientists are, they do not matter in the end—only Nature matters. . . . In the second version, however, we have a lot of work to do since, by analysing the allies and resources that settle a controversy we understand *everything* that there is to understand in technoscience. If the first version is correct, there is nothing for us to do apart from catching the most superficial aspects of science; if the second version is maintained, there is everything to understand except perhaps the most superfluous and flashy aspects of science. Given the stakes, the reader will realise why this problem should be tackled with caution. The whole book is in jeopardy here. (Latour 1987, p. 97, italics in the original)

Since "the whole book is in jeopardy here", let us look carefully at this passage. Latour says that if it is Nature that settles the controversies, the role of the sociologist is secondary, but if that is not the case, the sociologist can understand "*everything* that there is to understand in technoscience". How does he decide which version is the correct one? The answer appears in the subsequent text, where Latour distinguishes between the "cold parts of technoscience", for which "Nature is now taken as the cause of accurate descriptions of herself" (p. 100), and the active controversies, where Nature cannot be invoked:

When studying controversy—as we have so far—we cannot be *less* relativist than the very scientists and engineers we accompany; they do not *use* Nature as the external referee, and we have no reason to imagine that we are more clever than they are. (Latour 1987, p. 99, italics in the original)

In this quote and the previous one, Latour is playing constantly on the confusion between facts and our knowledge of them.[123] The correct answer to any scientific question, solved or

---

[123]An even more extreme example of this confusion appears in a recent article by Latour in *La Recherche*, a French monthly magazine devoted to the popularization of science (Latour 1998). Here Latour discusses what he interprets as the discovery in 1976, by French scientists working on the mummy of the pharaoh Ramses II, that his death (circa 1213 B.C.) was due to tuberculosis. Latour asks: "How could he pass

not, depends on the state of Nature (for example, on the number of neutrinos that the Sun really emits). Now, it happens that, for the unsolved problems, nobody knows the right answer, while for the solved ones, we do know it (at least if the accepted solution is correct, which can always be challenged). But there is no reason to adopt a "relativist" attitude in one case and a "realist" one in the other. The difference between these attitudes is a philosophical matter, and is independent of whether the problem is solved or not. For the relativist, there is simply no unique correct answer, independent of all social and cultural circumstances; this holds for the closed questions as well as for the open ones. On the other hand, the scientists who seek the correct solution are not relativist, almost by definition. Of course they *do* "use Nature as the external referee": that is, they seek to know what is really happening in Nature, and they design experiments for that purpose.

Let us not, however, leave the impression that the Third Rule of Method is *only* a triviality or a gross error. We'd like to give it one more interpretation (which is undoubtedly *not* Latour's own) that makes it at the same time interesting and correct. Let us read it as a methodological principle for a sociologist of science who does not himself have the scientific competence to make an independent assessment of whether the experimental/observational data do in fact warrant the con-

---

away due to a bacillus discovered by Robert Koch in 1882?" Latour notes, correctly, that it would be an anachronism to assert that Ramses II was killed by machine-gun fire or died from the stress provoked by a stock-market crash. But then, Latour wonders, why isn't death from tuberculosis likewise an anachronism? He goes so far as to assert that "Before Koch, the bacillus has no real existence." He dismisses the common-sense notion that Koch *discovered* a pre-existing bacillus as "having only the appearance of common sense". Of course, in the rest of the article, Latour gives no argument to justify these radical claims and provides no genuine alternative to the common-sense answer. He simply stresses the obvious fact that, in order to discover the cause of Ramses' death, a sophisticated analysis in Parisian laboratories was needed. But unless Latour is putting forward the truly radical claim that *nothing* we discover *ever* existed prior to its "discovery"—in particular, that no murderer is a murderer, in the sense that he committed a crime *before* the police "discovered" him to be a murderer—he needs to explain what is special about bacilli, and this he has utterly failed to do. The result is that Latour is saying nothing clear, and the article oscillates between extreme banalities and blatant falsehoods.

clusions the scientific community has drawn from them.[124] In such a situation, the sociologist will be understandably reluctant to say that "the scientific community under study came to conclusion X because X is the way the world really is"—*even if it is in fact the case* that X is the way the world is and that is the reason the scientists came to believe it—because the sociologist has no independent *grounds to believe* that X is the way the world really is other than the fact that the scientific community under study came to believe it. Of course, the sensible conclusion to draw from this *cul de sac* is that sociologists of science ought not to study scientific controversies on which they lack the competence to make an independent assessment of the facts, if there is no other (for example, historically later) scientific community on which they could justifiably rely for such an independent assessment. But it goes without saying that Latour would not enjoy this conclusion.[125]

Here lies, in fact, the fundamental problem for the sociologist of "science in action". It is not enough to study the alliances or power relationships between scientists, important though they may be. What appears to a sociologist as a pure power game may in fact be motivated by perfectly rational considerations which, however, can be understood as such only through a detailed understanding of the scientific theories and experiments.

Of course, nothing prevents a sociologist from acquiring such an understanding—or from working in collaboration with scientists who already have it—but in none of his Rules of Method does Latour recommend that sociologists of science follow this route. Indeed, in the case of Einstein's relativity, we

---

[124]The principle applies with particular force when such a sociologist is studying contemporary science, because in this case there is no other scientific community besides the one under study who could provide such an independent assessment. By contrast, for studies of the distant past, one can take advantage of what subsequent scientists learned, including the results from experiments going beyond those originally performed. See note 88 above.

[125]Nor would Steve Fuller, who asserts that "STS [Science and Technology Studies] practitioners employ methods that enable them to fathom both the 'inner workings' and the 'outer character' of science without having to be expert in the fields they study." (Fuller 1993, p. xii)

can show that Latour did not follow it himself.[126] This is understandable, because it is difficult to acquire the requisite knowledge, even for scientists working in a slightly different field. But
nothing is gained by biting off more than one can chew.

Practical Consequences

We don't want to give the impression that we are attacking only
some esoteric philosophical doctrines or the methodology followed by one current in the sociology of science. In fact, our target is much wider. Relativism (as well as other postmodern
ideas) has effects on the culture in general and on people's ways
of thinking. Here are a few examples we have come across. We
have no doubt that the reader will find many other examples in
the culture sections of newspapers, in certain educational theories, or simply in day-to-day conversations.

1. Relativism and criminal investigations. We have applied
various relativist arguments to criminal investigations in order
to show that, since they are thoroughly unconvincing in that
context, there is little reason to give them credence when applied to science. That is why the following excerpt is surprising,
to say the least: taken literally, it expresses a rather strong form
of relativism concerning precisely a criminal investigation. Here
is the context: In 1996, Belgium was shaken by a series of
kidnap-murders of children. In response to public outrage at
the inept police work, a parliamentary commission was set up
to examine the errors committed during the investigation. In a
spectacular televised session, two witnesses—a policeman
(Lesage) and a judge (Doutrèwe)—were confronted and ques-
tioned concerning the transmission of a key file. The policeman
swore he had sent the file to the judge, while the judge denied
having received it. The next day, an anthropologist of commu-

---

[126]See Chapter 6 below.

nication, Professor Yves Winkin of the University of Liège, was interviewed by one of the main Belgian newspapers (*Le Soir* of 20 December 1996):

> **Question:** The confrontation [between Lesage and Doutrèwe] was stimulated by an almost ultimate search for truth. Does truth exist?
> **Answer:** . . . I think that all the work of the commission is based on a sort of presupposition that there exists, not *a* truth, but *the* truth—which, if one presses hard enough, will finally come out.
>
> However, anthropologically, there are only partial truths, shared by a larger or smaller number of people: a group, a family, a firm. There is no transcendent truth. Therefore, I don't think that judge Doutrèwe or officer Lesage are hiding anything: both are telling their truth.
>
> Truth is always linked to an organization, depending upon the elements that are perceived as important. It is not surprising that these two people, representing two very different professional universes, should each set forth a different truth. Having said that, I think that, in this context of public responsibility, the commission can only proceed as it does.

This answer illustrates, in a striking way, the confusions into which some sectors of the social sciences have fallen through their use of a relativist vocabulary. The dispute between the policeman and the judge concerns, after all, a material fact: the transmission of a file. (It is, of course, possible that the file was sent but got lost on the way; but this remains a well-defined factual question.) Without a doubt, the epistemological problem is complicated: how is the commission to find out what really happened? Nevertheless, there *is* a truth of the matter: either the file was sent or it wasn't. It is hard to see what is gained by redefining the word "truth" (whether or not it is "partial") to mean simply "a belief shared by a larger or smaller number of people".

In this text, one also finds the idea of "different universes". Little by little, some tendencies in the social sciences have at-

omized humankind into cultures and groups having their own
conceptual universes—sometimes even their own "realities"—
and virtually unable to communicate with one another.[127] But in
this case it reaches a level bordering on the absurd: these two
people speak the same language, live less than a hundred miles
apart, and work in the criminal-justice system of a French-
speaking Belgian community comprising barely four million
people. Clearly, the problem does not arise from an inability to
communicate: the policeman and the judge understand perfectly well what is being asked, and they most likely know the
truth; quite simply, one of them has an interest in lying. But even
if they are both telling the truth—i.e., the file was sent but got
lost in transit, which is logically possible though unlikely—it
makes no sense to say that "both are telling *their* truth". Fortu-
nately, when it comes down to practical considerations, the anthropologist admits that the commission "can only proceed as
it does", that is, seek *the* truth. But what incredible confusions
before getting there.

2. Relativism and education. In a book written for high-
school teachers, whose goal is to explain "some notions of epistemology"[128], one finds the following definition:

> **Fact**
>   What one generally calls a fact is an interpretation of a sit-
> uation that no one, at least for the moment, wants to call into
> question. It should be remembered that, as the common lan-
> guage says, a fact becomes established, which illustrates well
> that we're talking about a theoretical model that one claims is
> appropriate.

---

[127]The so-called Sapir-Whorf thesis in linguistics appears to have played an important
role in this evolution: see note 37 above. Note also that Feyerabend, in his
autobiography (1995, pp. 151–152), disowned the radical-relativist use of the Sapir-
Whorf thesis that he had made in *Against Method* (Feyerabend 1975, chapter 17).

[128]The book's senior author is Gérard Fourez, a philosopher of science who is very
influential (at least in Belgium) in pedagogical matters, and whose book *La
Construction des sciences* (1992) has been translated into several languages.

Example: The assertions "The computer is on the desk" or "If one boils water, it evaporates" are considered to be factual propositions in the sense that no one wants to contest them at this moment in time. They are statements of theoretical interpretations that no one questions.

To assert that a proposition states a fact (that is, has the status of a factual or empirical proposition) is to claim that there is hardly any controversy about this interpretation at the moment one is speaking. But a fact can be put into question.

Example: For many centuries, it was considered to be a fact that the Sun revolves each day around the Earth. The appearance of another theory, such as that of the diurnal rotation of the Earth, entailed the replacement of the fact just cited by another: "The Earth rotates on its axis each day." (Fourez *et al.* 1997, pp. 76–77)

This confuses facts with *assertions* of fact.[129] For us, as for most people, a "fact" is a situation in the external world that exists irrespective of the knowledge we have (or don't have) of it—in particular, irrespective of any consensus or interpretation. Thus, it makes sense to say there are facts of which we are ignorant (Shakespeare's exact birth date, or the number of neutrinos emitted per second by the Sun). And there is a world of difference between saying that X killed Y and saying that no one, for the moment, wants to dispute this assertion (e.g., because X is black and everyone else is racist, or because biased news media successfully make people think that X killed Y). When it comes to a concrete example, the authors backtrack: they say that the Sun's revolution around the Earth was *considered to be* a fact, which amounts to admitting the distinction we are stressing (i.e., it was not *really* a fact). But in the next sentence they fall back into confusion: one fact has been replaced by another. Taken literally, in the *usual* sense of the word "fact", this would mean that the Earth has rotated on its axis only since Copernicus. But, of course, all the authors really

---

[129]Note that this appears in a text that is supposed to *enlighten* high-school teachers.

mean is that people's beliefs changed. Then why not say so, rather than confusing facts with (consensus) beliefs by using the same word to denote both concepts?[130]

A side benefit of the authors' non-standard notion of "fact" is that one can never be wrong (at least when asserting the same things as the people around us). A theory is never wrong in the sense that it is contradicted by the facts; rather, the facts change when the theories change.

Most importantly, it seems to us that a pedagogy based on this notion of "fact" is antithetical to encouraging a critical spirit in the student. In order to challenge prevailing assumptions— other people's as well as our own—it is essential to keep in mind that one *can* be wrong: that there exist facts independent of our claims, and that it is by comparison with these facts (to the extent we can ascertain them) that our claims have to be evaluated. When all is said and done, Fourez's redefinition of "fact" has—as Bertrand Russell noted in a similar context—all the advantages of theft over honest toil.[131]

3. Relativism in the Third World. Unfortunately, postmodern ideas are not confined to European philosophy departments or American literature departments. It seems to us that they do the most harm in the Third World, where the majority of the

---

[130]Or, worse, minimizing the importance of facts, not by giving any argument, but simply by ignoring them in favor of consensus beliefs. Indeed, the definitions in this book *systematically* conflate facts, information, objectivity, and rationality with—or reduce them to—intersubjective agreement. Moreover, a similar pattern is found in Fourez's *La Construction des sciences* (1992). For example (p. 37): "To be 'objective' means to follow instituted rules. . . . Being 'objective' is not the opposite of being 'subjective': rather, it is to be subjective in a certain way. But it is not to be individually subjective since one will follow socially instituted rules . . .". This is highly misleading: following rules does not ensure objectivity in the usual sense (people who blindly repeat religious or political slogans certainly follow "socially instituted rules", but they can hardly be called objective) and people can be objective while breaking many rules (e.g., Galileo).

[131]Note also that defining "fact" as "there is hardly any controversy . . ." runs into a logical problem: Is the absence of controversy itself a fact? And if so, how to define it? By the absence of controversy about the assertion that there is no controversy? Obviously, Fourez and his colleagues are using in the social sciences a naively realist epistemology that they implicitly reject for the natural sciences. See p. 83–84 above for an analogous inconsistency in Feyerabend.

world's population lives and where the supposedly "passé" work of the Enlightenment is far from complete.

Meera Nanda, an Indian biochemist who used to work in the "Science for the People" movements in India and who is now studying sociology of science in the United States, tells the following story about the traditional Vedic superstitions governing the construction of sacred buildings, which aim at maximizing "positive energy". An Indian politician, who found himself in hot water, was advised that

> his troubles would vanish if he entered his office from an east-facing gate. But on the east side of his office there was a slum through which his car could not pass. [So he] ordered the slum to be demolished. (Nanda 1997, p. 82)

Nanda observes, quite rightly, that

> If the Indian left were as active in the people's science movement as it used to be, it would have led an agitation not only against the demolition of people's homes, but also against the superstition that was used to justify it. . . . A left movement that was not so busy establishing "respect" for non-Western knowledge would never have allowed the power-wielders to hide behind indigenous "experts."
>
> I tried out this case on my social constructionist friends here in the United States. . . . [They told me] that seeing the two culturally bound descriptions of space[132] at par with each other is progressive in itself, for then *neither* can claim to know the absolute truth, and thus tradition will lose its hold on people's minds. (Nanda 1997, p. 82)

The problem with this kind of answer is that practical choices have to be made—what type of medicine to use, or in which direction to orient buildings—and at this point theoretical non-

---

[132]That is, the scientific view and the one based on traditional Vedic ideas. [Note added by us.]

chalance becomes untenable. As a result, intellectuals easily fall into the hypocrisy of using "Western" science when it is essential—for example, when they are *seriously* ill—while urging the common people to put their faith in superstitions.

5. Luce Irigaray

Luce Irigaray's writings have dealt with a wide variety of topics, ranging from psychoanalysis to linguistics to the philosophy of science. In this latter field, she maintains that

> Every piece of knowledge is produced by subjects in a given historical context. Even if that knowledge aims to be objective, even if its techniques are designed to ensure objectivity, science always displays certain choices, certain exclusions, and these are particularly determined by the sex of the scholars involved. (Irigaray 1993, p. 204)

In our opinion this thesis deserves an in-depth study. Let us look, however, at the examples Irigaray gives to illustrate it in the physical sciences:

> This [scientific] subject today is enormously interested in acceleration that goes beyond our human powers, in weightlessness, in crossing through natural space and time, in overcoming cosmic rhythms and their regulation. He is also interested in disintegration, fission, explosion, catastrophes, etc. This reality can be confirmed from within the natural and the human sciences. (Irigaray 1993, p. 204)

This catalogue of contemporary scientific work is rather arbitrary, and quite vague: what is the meaning of "acceleration that goes beyond our human powers", "crossing through natural space and time", or "overcoming cosmic rhythms and their regulation"? But what follows is even stranger:

– If the identity of the human subject is defined in the work of Freud by a *Spaltung*, this is also the word used for nuclear fission. Nietzsche also perceived his ego as an atomic nucleus threatened with explosion. As for Einstein, the main issue he raises, in my mind, is that, given his interest in accelerations without electromagnetic reequilibrations, he leaves us with only one hope, his God. It is true that Einstein played the violin: music helped him preserve his personal equilibrium. But what does the mighty theory of general relativity do for us except establish nuclear power plants and question our bodily inertia, that necessary condition of life?

– As for the astronomers, Reaves, following up on the American big bang theory, describes the origin of the universe as an explosion. How is it that this current interpretation so closely parallels the abstracts of the whole field of other scientific discoveries?

– René Thom, another theoretician who works at the intersection of science and philosophy, talks about catastrophes through conflicts rather than about generation through abundance, growth, positive attraction, particularly in nature.

– Quantum mechanics is interested in the disappearance of the world.

– Scientists today are working on smaller and smaller particles, which cannot be perceived but only defined thanks to sophisticated technical instruments and bundles of energy. (Irigaray 1993, pp. 204–205)

Let us consider these arguments one by one:

– About the *Spaltung*, Irigaray's "logic" is truly bizarre: Does she really think that this linguistic coincidence constitutes an argument? And if so, what does it show?

– Concerning Nietzsche: the atomic nucleus was discovered in 1911, and nuclear fission in 1938; the possibility of a nuclear chain reaction, leading to an explosion, was studied theoretically during the late 1930s and sadly realized experimentally during the 1940s. It is thus highly improbable that Nietzsche (1844–1900) could have perceived his ego "as an atomic nucleus threatened with explosion". (Of course, this hasn't the slightest importance: even if Irigaray's claim about Nietzsche were correct, what would it imply?)

– The expression "accelerations without electromagnetic reequilibrations" has no meaning in physics; it is entirely Irigaray's invention. It goes without saying that Einstein could not possibly have been interested in this nonexistent subject.

– General relativity bears no relation to nuclear power plants; Irigaray has probably confused it with special relativity, which does apply to nuclear power plants as well as to many other things (elementary particles, atoms, stars . . .). The concept of inertia certainly appears in relativity theory, as it does in Newtonian mechanics; but it has nothing to do with human beings' "bodily inertia", whatever this is intended to mean.[133]

– In what way does the cosmological theory of the Big Bang "so closely parallel . . . the whole field of other scientific discoveries"? Which other discoveries, made at what time? Irigaray doesn't say. The bottom line is that the Big Bang theory, which dates back to the late 1920s, is today supported by a plethora of astronomical observations.[134]

---

[133]For good introductions to special and general relativity, see, for example, Einstein (1960 [1920]), Mermin (1989), and Sartori (1996).

[134]During the 1920s, the astronomer Edwin Hubble discovered that the galaxies are moving away from the Earth, at speeds that are proportional to their distance from the Earth. Between 1927 and 1931, various physicists proposed explanations of this expansion within the framework of Einstein's general relativity (without making the Earth a privileged center of observation) as arising from an initial cosmic "explosion"; this theory was later nicknamed the "Big Bang". But, though the Big Bang hypothesis explains the observed expansion in a very natural way, it is not the only possible theory: towards the end of the 1940s, the astrophysicists Hoyle, Bondi, and Gold proposed the alternative theory of the "Steady State Universe", according to which there is a general expansion *without* a primeval explosion (but with the continuous creation of new matter). However, in 1965, the physicists Penzias and Wilson discovered (by accident!) the cosmic microwave background radiation, whose spectrum and almost-isotropy turned out to be in complete agreement with the prediction based on general relativity for a "residue" from the Big Bang. In part because of this observation, but also for many other reasons, the Big Bang theory is today almost universally accepted among astrophysicists, though there is a lively debate on the details. For a nontechnical introduction to the Big Bang theory and the observational data supporting it, see Weinberg (1977), Silk (1989), and Rees (1997).

The "Reaves" to whom Irigaray refers is presumably Hubert Reeves, a Canadian astrophysicist living in France who has written several popular books on cosmology and astrophysics.

– It is true that in some (highly debatable) interpretations of quantum mechanics, the concept of objective reality at the atomic level is called into question, but this has nothing to do with "the disappearance of the world". Perhaps Irigaray is alluding to cosmological theories about the end of the universe (the "Big Crunch"), but quantum mechanics does not play a major role in these theories.[135]

– Irigaray correctly observes that subatomic physics deals with particles that are too small to be directly perceived by our senses. But it is hard to see how this relates to the gender of the researchers. Is the use of instruments to extend the range of human sense perceptions a particularly "masculine" characteristic? Marie Curie and Rosalind Franklin might beg to differ.

Let us consider, finally, an argument put forward elsewhere by Irigaray:

> Is $E = Mc^2$ a sexed equation? Perhaps it is. Let us make the hypothesis that it is insofar as it privileges the speed of light over other speeds that are vitally necessary to us. What seems to me to indicate the possibly sexed nature of the equation is not directly its uses by nuclear weapons, rather it is having privileged what goes the fastest . . . (Irigaray 1987b, p. 110)

Whatever one may think about the "other speeds that are vitally necessary to us", the fact remains that the relationship $E = Mc^2$ between energy (E) and mass (M) is experimentally verified to a high degree of precision, and it would obviously not be valid if the speed of light (c) were replaced by another speed.

In summary, it seems to us that the influence of cultural, ideological, and sexual factors on scientific choices—the subjects studied, the theories put forward—is an important research topic in the history of science and deserves a rigorous

---

[135]Except in the last millionth of a billionth of a billionth of a billionth of a billionth of a second, when quantum gravitational effects become important.

investigation. But, to contribute usefully to this research, one must understand at a rather deep level the scientific fields under analysis. Unfortunately, Irigaray's claims show a superficial understanding of the subjects she addresses, and consequently bring nothing to the discussion.

Fluid Mechanics

Some years earlier, in an essay entitled "The 'Mechanics' of Fluids", Irigaray had already elaborated her critique of "masculine" physics: she seems to claim that fluid mechanics is underdeveloped relative to solid mechanics because solidity is identified (according to her) with men and fluidity with women. (But Irigaray was born in Belgium: doesn't she know the symbol of the city of Brussels?) One of Irigaray's American interpreters summarizes her argument as follows:

> The privileging of solid over fluid mechanics, and indeed the inability of science to deal with turbulent flow at all, she attributes to the association of fluidity with femininity. Whereas men have sex organs that protrude and become rigid, women have openings that leak menstrual blood and vaginal fluids. Although men, too, flow on occasion—when semen is emitted, for example—this aspect of their sexuality is not emphasized. It is the rigidity of the male organ that counts, not its complicity in fluid flow. These idealizations are reinscribed in mathematics, which conceives of fluids as laminated planes and other modified solid forms. In the same way that women are erased within masculinist theories and language, existing only as not-men, so fluids have been erased from science, existing only as not-solids. From this perspective it is no wonder that science has not been able to arrive at a successful model for turbulence. The problem of turbulent flow cannot be solved because the conceptions of fluids (and of women) have been formulated so as necessarily to leave unarticulated remainders. (Hayles 1992, p. 17)

It seems to us that Hayles' exegesis of Irigaray's ideas is much clearer than the original. Nevertheless, because of the obscurity of Irigaray's text, we cannot guarantee that Hayles has faithfully explicated Irigaray's meaning. Hayles, for her part, rejects Irigaray's reasoning on the grounds that it is too distant from the scientific facts (see note 137 below), but tries to arrive at similar conclusions by a different route. In our opinion, Hayles' attempt doesn't fare much better than Irigaray's, but at least it is expressed more clearly.[136]

Let us now try to follow the details of Irigaray's argument. Her essay begins as follows:

> It is already getting around—at what rate? in what contexts? in spite of what resistances?—that women diffuse themselves according to modalities scarcely compatible with the framework of the ruling symbolics. Which doesn't happen without causing some turbulence, we might even say some whirlwinds, that ought to be reconfined within solid walls of

---

[136]Hayles' argument begins with an explanation of the important conceptual differences between linear differential equations and the nonlinear ones arising in fluid mechanics. It's a respectable attempt at scientific journalism, albeit marred by a few errors (e.g. she confuses feedback with nonlinearity, and she asserts that Euler's equation is linear). From this point on, however, her argument deteriorates into a caricature of postmodern lit-crit. Seeking to trace the historical development of fluid mechanics in the period 1650–1750, she claims to identify "a pair of hierarchical dichotomies [what else?!] in which the first term is privileged at the expense of the second: continuity versus rupture, and conservation versus dissipation." (Hayles 1992, p. 22) There follows a rather confused discussion of the conceptual foundations of differential calculus, an imaginative (to say the least) exegesis of the "subliminal gender identifications" in early hydraulics, and a Freudian analysis of thermodynamics "from heat death to *jouissance*". Hayles concludes by asserting a radically relativist thesis:

Despite their names, conservation laws are not inevitable facts of nature but constructions that foreground some experiences and marginalize others. . . . Almost without exception, conservation laws were formulated, developed, and experimentally tested by men. If conservation laws represent particular emphases and not inevitable facts, then people living in different kinds of bodies and identifying with different gender constructions might well have arrived at different models for flow. (Hayles 1992, pp. 31–32)

But she gives no argument to support her claim that the laws of conservation of energy and momentum, for example, might be other than "inevitable facts of nature"; nor does she give the slightest hint of *what* kinds of "different models for flow" might have been arrived at by "people living in different kinds of bodies".

principle, to keep them from spreading to infinity. Otherwise they might even go so far as to disturb that third agency designated as the real—a transgression and confusion of boundaries that it is important to restore to their proper order.

So we shall have to turn back to "science" in order to ask it some questions. [**Footnote:** The reader is advised to consult some texts on solid and fluid mechanics.[137]] Ask, for example, about its *historical lag in elaborating a "theory" of fluids*, and about the ensuing aporia even in mathematical formalization. A postponed reckoning that was eventually to be imputed to the real. [**Footnote:** Cf. the signification of the "real" in the writings of Jacques Lacan *(Écrits, Séminaires).*]

Now if we examine the properties of fluids, we note that this "real" may well include, and in large measure, *a physical reality* that continues to resist adequate symbolization and/or that signifies the powerlessness of logic to incorporate in its writing all the characteristic features of nature. And it has often been found necessary to minimize certain of these features of nature, to envisage them, and it, only in light of an ideal status, so as to keep it/them from jamming the works of the theoretical machine.

But what division is being perpetuated here between a language that is always subject to the postulates of ideality and an empirics that has forfeited all symbolization? And how can we fail to recognize that with respect to this caesura, to the schism that underwrites the purity of logic, language remains necessarily meta-"something"? Not simply in its artic-

---

[137]Hayles, who is in general favorable to Irigaray, notes that:

> From talking with several applied mathematicians and fluid mechanicists about Irigaray's claim, I can testify that they unanimously conclude she does not know the first thing about their disciplines. In their view, her argument is not to be taken seriously.
>
> There is evidence to support this view. In a footnote to the chapter's first page, Irigaray airily advises the reader "to consult some texts on solid and fluid mechanics" without bothering to mention any. The lack of mathematical detail in her argument forces one to wonder whether she has followed this advice herself. Nowhere does she mention a name or date that would enable one to connect her argument with a specific theory of fluids, much less to trace debates between opposing theories. (Hayles 1992, p. 17)

ulation, in its utterance, here and now, by a subject, but because, owing to his own structure and unbeknownst to him,
that "subject" is already repeating normative "judgments" on
a nature that is resistant to such a transcription.

And how are we to prevent the very unconscious (of the)
"subject" from being prorogated as such, indeed diminished in
its interpretation, by a systematics that remarks [*sic*] a historical "inattention" to fluids? In other words: what structuration of (the) language does not maintain a *complicity of
long standing between rationality and a mechanics of solids
alone?* (Irigaray 1985a, pp. 106–107)

Irigaray's claims about solid and fluid mechanics demand some
comment. First of all, solid mechanics is far from being complete; it has many unsolved problems, such as the quantitative
description of fractures. Secondly, fluids in equilibrium or in
laminar flow are relatively well understood. Besides, we know
the equations—the so-called Navier-Stokes equations—that
govern the behavior of fluids in a vast number of situations. The
main problem is that these nonlinear partial differential equations are very difficult to solve, in particular for turbulent
flows.[138] But this difficulty has nothing to do with any "powerlessness of logic" or failure of "adequate symbolization", nor
with the "structuration of (the) language". Here Irigaray follows
her (ex-)teacher Lacan, in insisting too much on the logical formalism at the expense of the physical content.

Irigaray continues with a bizarre mélange of fluids, psychoanalysis, and mathematical logic:

> Certainly the emphasis has increasingly shifted from the
> definition of terms to the analysis of relations among terms
> (Frege's theory is one example among many). This has even
> led to the recognition of *a semantics of incomplete beings*:
> functional symbols.

---

[138]For a non-technical explanation of the concept of linearity (applied to an equation),
see p. 143 below.

But, beyond the fact that the indeterminacy thus allowed in the proposition is subject to a general implication of the *formal* type—the variable is such only within the limits of the identity of (the) form(s) of syntax—a preponderant role is left to the *symbol of universality*—to the universal quantifier—whose modalities of recourse to the geometric still have to be examined.

Thus the "all"—of x, but also of the system—has already prescribed the "not-all" of each particular relation established, and that "all" is such only by a definition of extension that cannot get along without projection onto a given space-map, whose between(s) will be given their value(s) on the basis of punctual frames of reference.[139]

The "place" thus turns out to have been in some way planned and punctuated for the purpose of calculating each "all," but also the "all" of the system. Unless it is allowed to extend to infinity, which rules out in advance any determination of value for either the variables or their relations.

But where does that place—of discourse—find its "greater-than-all" in order to be able to form(alize) itself in this way? To systematize itself? And won't that greater than "all" come back from its denegation—from its forclusion?— in modes that are still theo-logical [*sic*]? Whose relation to the feminine "not-all" remains to be articulated: *God or feminine pleasure.*

While she waits for these divine rediscoveries, awoman [*sic*] serves (only) as a *projective map* for the purpose of guaranteeing the totality of the system—the excess factor of its "greater than all"; she serves as *a geometric prop* for evaluating the "all" of the extension of each of its "concepts" including those that are still undetermined, serves as fixed and congealed *intervals* between their definitions in "language,"

---

[139]The three preceding paragraphs, which supposedly concern mathematical logic, are devoid of meaning, with one exception: the assertion that "a preponderant role is left to . . . the universal quantifier" is meaningful and false (see note 143 below).

and as the possibility of *establishing individual relation-ships* among these concepts. (Irigaray 1985a, pp. 107–108)

A bit farther down, Irigaray returns to fluid mechanics:

> What is left uninterpreted in the economy of fluids—the resistances brought to bear upon solids, for example—is in the end given over to God. Overlooking the properties of *real* fluids—internal frictions, pressures, movements, and so on, that is, *their specific dynamics*—leads to giving the real back to God, as only the idealizable characteristics of fluids are included in their mathematicization.
>
> Or again: considerations *of* pure mathematics have precluded the analysis of fluids except in terms of laminated planes, solenoid movements (of a current privileging the relation to an axis), spring-points, well-points, whirlwind-points, which have only an approximate relation to reality. Leaving some *remainder.* Up to *infinity*: the center of these "movements" corresponding to zero supposes in them an infinite speed, which is *physically unacceptable.* Certainly these "theoretical" fluids have enabled the technical—also mathematical—form of analysis to progress, while losing a certain relationship to *the reality of bodies in the process.*
>
> *What consequences does this have for "science" and psy-choanalytic practice?* (Irigaray 1985a, p. 109)

In this passage, Irigaray shows that she does not understand the role of approximations and idealizations in science. First of all, the Navier-Stokes equations are approximations that are valid only on a macroscopic (or at least supra-atomic) scale, because they treat the fluid as a continuum and neglect its molecular structure. And since these equations are themselves very hard to solve, mathematicians try to study them first in idealized situations or through more-or-less controlled approximations. But the fact that, for example, the speed is infinite at the center of a vortex means only that the approximation ought not be taken too seriously near that point—as was obvious from the start, since the approach is in any case valid only on scales much larger than that of molecules. In any case, nothing is "given over to God"; there are, quite simply, scientific problems left for future generations.

Finally, it is hard to see what relation, besides a purely metaphorical one, fluid mechanics could have with psychoanalysis. Suppose that tomorrow someone were to come up with a satisfactory theory of turbulence. In what way would (or should) that affect our theories of human psychology?

One could continue quoting Irigaray, but the reader is probably lost (so are we). She concludes her essay with some words of consolation:

> And if, by chance, you were to have the impression of not having yet understood everything, then perhaps you would do well to leave your ears half-open for what is in such close touch with itself that it confounds your discretion. (Irigaray 1985a, p. 118)

All in all, Irigaray fails to understand the nature of the physical and mathematical problems arising in fluid mechanics. Her discourse is based solely on vague analogies that, moreover, mix up the theory of real fluids with its already analogical use in psychoanalysis. Irigaray seems to be aware of this problem, as she answers it as follows:

> And if anyone objects that the question, put this way, relies too heavily on metaphors, it is easy to reply that the question in fact impugns the privilege granted to metaphor (a quasi solid) over metonymy (which is much more closely allied to fluids). (Irigaray 1985a, pp. 109–110)

Alas, this reply reminds us of the old Jewish joke: "Why does a Jew always answer a question with a question?" "And why shouldn't a Jew answer a question with a question?"

Mathematics and Logic

As we have seen, Irigaray has a penchant for reducing problems in the physical sciences to games of mathematical formalization or even of language. Unfortunately, her knowledge of mathematical logic is as superficial as her knowledge of physics. An illustration of this can be found in her famous essay "Is the Subject of Science Sexed?" After a rather idiosyncratic sketch of the scientific method, Irigaray goes on to write:

> These characteristics reveal an isomorphism in man's sexual Imaginary, an isomorphism which must remain rigorously masked. "Our subjective experiences and our beliefs can never justify any utterance," affirms the epistemologist of the sciences.
>
> You must add that all of these discoveries must be expressed in a language that is *well-written*, meaning *reasonable*, that is:
>
> – expressed in symbols or letters, interchangeable with *proper nouns*, that refer only to an intra-theoretical object, thus to no character or object from the real or from reality. The scholar enters into a fictional universe that is incomprehensible to those who do not participate in it.
>
> (Irigaray 1985b, p. 312; Irigaray 1987a, p. 73)

Here again one encounters Irigaray's misunderstandings concerning the role of mathematical formalism in science. It is not true that all the concepts of a scientific theory "refer only to an intra-theoretical object". Quite the contrary, at least *some* of the theory's concepts must correspond to something in the real world, for otherwise the theory would have no empirical consequences whatsoever (and thus not be scientific). Consequently, the scientist's universe is not populated solely by fictions. Finally, neither the real world nor the scientific theories that explain it are completely incomprehensible to non-experts; in many cases, there exist good popular or semi-popular expositions.

The remainder of Irigaray's text is both pedantic and unwittingly comic:

> — the formative signs for terms and for predicates are:
>   + : or definition of a new term[140];
>   = : which indicates a property by equivalence and substitution (belonging to a whole or to a world);
>   ∈ : signifying belonging to an object type
> — the *quantifiers* (and not *qualifiers*) are:
>   ≥ ≤;
>   the universal quantifier;
>   the existential quantifier submitted, as its name indicates, to the quantitative.
>
>   According to the semantics of incomplete beings (Frege), functional symbols are variables found at the boundary of the identity of syntactic forms and the dominant role is given to the universality symbol or universal quantifier.
> — the *connectors* are:
>   - negation: P or not P[141];
>   - conjunction: P or Q[142];
>   - disjunction: P or Q;
>   - implication: P implies Q;
>   - equivalence: P equals Q;
>   There is then no sign:
>   - of *difference* other than the quantitative;

---

[140]As we all learned in elementary school, the symbol "+" denotes the addition of two numbers. We are at a loss to explain how Irigaray got the idea that it indicates the "definition of a new term".

[141]We apologize to the reader for our pedantry: the negation of a proposition P is not "P or not P", but simply "not P".

[142]This is presumably a typographical error; it occurs also in the original French and was overlooked by the translator. The conjunction of two propositions is, of course, "P *and* Q".

- of *reciprocity* (other than within a common
   property or a common whole);
- of *exchange;*
- of *fluidity.*
(Irigaray 1985b, pp. 312–313; Irigaray 1987a, pp. 73–74)

To begin with, Irigaray has confused the concept of "quantification" in logic with the word's everyday meaning (i.e., making something quantitative or numerical); in actual fact, there is no relationship between these two concepts. The quantifiers in logic are "for all" (universal quantifier) and "there exists" (existential quantifier). For example, the sentence "x likes chocolate" is a statement about a certain individual x; the universal quantifier transforms it into the statement "for all x [belonging to some set assumed known], x likes chocolate", while the existential quantifier transforms it into "there exists at least one x [belonging to some set assumed known] such that x likes chocolate". This clearly has nothing to do with numbers, and Irigaray's purported opposition between "quantifiers" and "qualifiers" is meaningless.

Besides, the inequality signs "$\geq$" (greater than or equal to) and "$\leq$" (less than or equal to) are not quantifiers. They relate to quantification in the ordinary sense of the word, not in the sense of quantifiers in logic.

Moreover, no "dominant role" is granted to the universal quantifier. Quite the contrary, there is a perfect symmetry between the universal and existential quantifiers, and any proposition using one of them can be transformed into a logically equivalent proposition using the other (at least in classical logic, which is Irigaray's supposed subject).[143] This is an elementary fact, taught in every introductory logic course; it is surprising that Irigaray, who speaks so much about mathematical logic, should fail to know it.

---

[143] To see this, let $P(x)$ be any statement about an individual x. The proposition "for all x, $P(x)$" is equivalent to "there does not exist x such that $P(x)$ is false". Similarly, the proposition "there exists at least one x such that $P(x)$" is equivalent to "it is false that, for all x, $P(x)$ is false".

Finally, her assertion that there is no sign (or, what is more relevant, no *concept*) of difference other than the quantitative is false. In mathematics, there are many types of objects other than numbers—for example, sets, functions, groups, topological spaces, etc.—and, when talking about two such objects, one may of course say that they are identical or different. The standard equality sign (=) is used to indicate that they are identical, and the standard inequality sign (≠) to indicate that they are different.

Later in the same essay, Irigaray claims to unmask also the sexist biases at the heart of "pure" mathematics:

> - the *mathematical sciences*, in the theory of wholes [*théorie des ensembles*], concern themselves with closed and open spaces, with the infinitely big and the infinitely small.[144] They concern themselves very little with the question of the partially open, with wholes that are not clearly delineated [*ensembles flous*], with any analysis of the problem of borders [*bords*], of the passage between, of fluctuations occurring between the thresholds of specific wholes. Even if topology suggests these questions, it emphasizes what closes rather than what resists all circularity. (Irigaray 1985b, p. 315; Irigaray 1987a, pp. 76–77[145])

Irigaray's phrases are vague: "the partially open", "the passage between", "fluctuations between the thresholds of specific wholes"—what exactly is she talking about? For what it's worth, the "problem" of boundaries [*bords*], far from being neglected, has been at the center of algebraic topology since its inception a century ago[146], and "manifolds with boundary" [*var-*

---

[144]In actual fact, set theory *(théorie des ensembles)* studies the properties of "bare" sets, that is, sets without any topological or geometrical structure. The questions alluded to here by Irigaray belong rather to topology, geometry, and analysis.

[145]Let us remark that the published English translation, quoted above, contains several errors. *Théorie des ensembles* is "set theory", not "theory of wholes". *Ensembles flous* presumably refers to the mathematical theory of "fuzzy sets". *Bords* is best translated in the mathematical context as "boundaries".

[146]See, for example, Dieudonné (1989).

iétés à bord] have been actively studied in differential geometry for at least fifty years. And, last but not least, what does all this have to do with feminism?

We were therefore quite surprised to find this passage quoted in a recent book devoted to the teaching of mathematics. The author is a prominent American feminist pedagogue of mathematics, whose goal—which we share wholeheartedly— is to attract more young women to scientific careers. She quotes approvingly this text of Irigaray and continues by saying:

> In the context provided by Irigaray we can see an opposition between the linear time of mathematics problems of related rates, distance formulas, and linear acceleration versus the dominant experiential cyclical time of the menstrual body. Is it obvious to the female mind-body that intervals have end-points, that parabolas neatly divide the plane, and, indeed, that the linear mathematics of schooling describes the world of experience in intuitively obvious ways?[147] (Damarin 1995, p. 252)

This theory is startling, to say the least: Does the author really believe that menstruation makes it more difficult for young women to understand elementary notions of geometry? This view is uncannily reminiscent of the Victorian gentlemen who held that women, with their delicate reproductive organs, are unsuited to rational thought and to science. With friends like these, the feminist cause hardly needs enemies.[148]

---

[147]Let us remark that, in this passage, the word "linear" is used three times, inappropriately and with three different apparent intended meanings. See p. 143–45 below for a discussion of abuses of the word "linear".

[148]Nor is this an isolated case. Hayles concludes her article on fluid mechanics by saying that

> the experiences articulated in this essay are shaped by the struggle to remain within the bounds of rational discourse while still questioning some of its major premises. Whereas the flow of the argument has been female and feminist, the channel into which it has been directed is male and masculinist. (Hayles 1992, p. 40)

Hayles thus appears to accept, without the slightest hint of self-consciousness, the identification of "rational discourse" with "male and masculinist".

One finds similar ideas in Irigaray's own writings. Indeed, her scientific confusions are linked to, and are taken to provide support for, more general philosophical considerations of a vaguely relativist nature. Starting from the idea that science is "masculine", Irigaray rejects "the belief in a truth independent of the subject" [*la croyance en une vérité indépendante du sujet*] and advises women not to

> accept or subscribe to the existence of a neutral, universal science, to which women should painfully gain access and with which they then torture themselves and taunt other women, transforming science into a new superego. (Irigaray 1993, p. 203)

These claims are clearly very debatable. To be sure, they are accompanied by more nuanced assertions, for example: "Truth is always the product of some man or woman. This does not mean that truth contains no objectivity"; and "All truth is partially relative."[149] The problem is to know exactly what Irigaray wants to say and how she intends to resolve these contradictions.

The roots of the tree of science may be bitter, but its fruits are sweet. To say that women should shun a universal science amounts to infantilizing them. To link rationality and objectivity to the male, and emotion and subjectivity to the female, is to repeat the most blatant sexist stereotypes. Speaking of the female "sexual economy" from puberty through menopause, Irigaray writes:

> But every stage in this development has its own temporality, which is possibly cyclic and linked to cosmic rhythms. If women have felt so terribly threatened by the accident at

---

[149]Irigaray (1993, p. 203).

Chernobyl, that is because of the irreducible relation of their bodies to the universe. (Irigaray 1993, p. 200)[150]

Here Irigaray falls straight into mysticism. Cosmic rhythms, relation to the universe—what on earth is she talking about? To reduce women to their sexuality, menstrual cycles, and rhythms (cosmic or not) is to attack everything the feminist movement has fought for during the last three decades. Simone de Beauvoir must be turning in her grave.

---

[150]For some even more shocking statements in the same vein, see Irigaray (1987b, pp. 106–108).

6. Bruno Latour

The sociologist of science Bruno Latour is well-known for his book *Science in Action*, which we have briefly analyzed in Chapter 4. Much less well known, however, is his semiotic analysis of the theory of relativity, in which "Einstein's text is read as a contribution to the sociology of delegation" (Latour 1988, p. 3). In this chapter, we shall examine Latour's interpretation of relativity and show that it illustrates perfectly the problems encountered by a sociologist who aims to analyze the content of a scientific theory he does not understand very well.

Latour considers his article as a contribution to, and extension of, the strong programme in the sociology of science, which asserts that "the content of any science is social through and through" (p. 3). According to Latour, the strong programme has had "some degree of success in the empirical sciences" but less in the mathematical sciences (p. 3). He complains that previous social analyses of Einstein's theory of relativity have "shunned the technical aspects of his theory" and failed to give any "indication of how relativity theory *itself* could be said to be social" (pp. 4–5, italics in the original). Latour sets himself the ambitious task of vindicating this latter idea, which he proposes to do by redefining the concept of "social" (pp. 4–5). For the sake of brevity, we won't enter into the sociological conclusions Latour purports to draw from his study of relativity, but shall simply point out that his argument is undermined by several fundamental misunderstandings about the theory of relativity itself.[151]

---

[151]Let us nevertheless quote the physicist Huth (1998), who has also made a critical analysis of Latour's article: "In this article, the meanings of the terms 'society' and

Latour bases his analysis of the theory of relativity on a semiotic reading of Einstein's semi-popular book, *Relativity: The Special and the General Theory* (1920). After a survey of semiotic notions such as "shifting in" and "shifting out" of narrators, Latour attempts to apply these notions to Einstein's special theory of relativity. But, in so doing, he misunderstands the concept of "frame of reference" in physics. A brief digression is therefore in order.

In physics, a *frame of reference* is a scheme for assigning spatial and temporal coordinates (x,y,z,t) to "events". For example, an event in New York City can be located by saying that it takes place at the corner of 6th Avenue (x) and 42nd Street (y), 30 meters above ground level (z), at noon on May 1, 1998 (t). In general, a frame of reference can be visualized as a rigid rectangular framework of meter sticks and clocks, which together allow coordinates of "where" and "when" to be assigned to any event.

Obviously, setting up a frame of reference involves making a number of arbitrary choices: for example, where to locate the origin of spatial coordinates (here 0th Avenue and 0th Street at ground level), how to orient the spatial axes (here east-west, north-south, up-down), and where to locate the origin of time (here midnight on January 1, year 0). But this arbitrariness is relatively trivial, in the sense that if we make any other choice of origins and orientations, there are quite simple formulae for translating from the former coordinates to the latter.

A more interesting situation arises when we consider two frames of reference in relative *motion*. For example, one frame of reference might be attached to the Earth, while another is attached to a car moving at 100 meters per second eastwards relative to the Earth. Much of the history of modern physics since Galileo concerns the question of whether the laws of physics take the same form with respect to each of these two frames of reference, and what equations are to be used for translating

'abstraction' are so stretched to fit his interpretation of relativity that they lose any semblance of common meaning, and shed no new light on the theory itself."

from the former coordinates (x,y,z,t) to the latter (x',y',z',t'). In particular, Einstein's theory of relativity deals precisely with these two questions.[152]

In pedagogical presentations of the theory of relativity, a frame of reference is often equated loosely with an "observer". More precisely, a frame of reference can be identified with a *set* of observers, one placed at *each* point in space, all at rest with respect to one another, and all equipped with suitably synchronized clocks. But it is crucial to note that these "observers" need not be humans: a frame of reference can perfectly well be constructed entirely out of machines (as is nowadays done routinely in high-energy-physics experiments). Indeed, a frame of reference need not be "constructed" at all: it makes perfect sense to imagine the frame of reference attached to a moving proton in a high-energy collision.[153]

Returning to Latour's text, we may distinguish three errors in his analysis. First of all, he appears to think that relativity is concerned with the relative *location* (rather than the relative *motion*) of different frames of reference, at least in the following excerpts:

I will use the following diagram in which the two (or more) frames of reference mark different *positions* in space and time . . . (p. 6)

[N]o matter how *far away* I delegate the observers, they all send back superimposable reports . . . (p. 14)

[E]ither we maintain absolute space and time and the laws of nature become different in different *places* . . . (p. 24)

---

[152]For a good introduction to the theory of relativity, see, for example, Einstein (1960 [1920]), Mermin (1989), or Sartori (1996).

[153]Indeed, by interpreting the collision of two protons with respect to the frame of reference attached to one of them, one can learn important things about the internal structure of protons.

[P]rovided the two relativities [special and general] are accepted, more frames of reference with less privilege can be accessed, reduced, accumulated and combined, observers can be delegated to a few more *places* in the infinitely large (the cosmos) and the infinitely small (electrons), and the readings they send will be understandable. His [Einstein's] book could well be titled: 'New Instructions for Bringing Back Long-Distance Scientific Travellers'. (pp. 22–23)

(Italics added)

This error can perhaps be attributed to a lack of precision in Latour's style. A second error—which is in our opinion more serious, and is indirectly related to the first—comes from an apparent confusion between the concept of "frame of reference" in physics and that of "actor" in semiotics:

How can one decide whether an observation made in a train about the behaviour of a falling stone can be made to coincide with the observation made of the same falling stone from the embankment? If there are only one, or even *two*, frames of reference, no solution can be found . . . Einstein's solution is to consider *three* actors: one in the train, one on the embankment and a third one, the author [enunciator] or one of its representants, who tries to superimpose the coded observations sent back by the two others. (pp. 10–11, italics in the original)

In reality, Einstein never considers three frames of reference. The Lorentz transformations[154] allow one to establish a correspondence between the coordinates of an event in *two* frames of reference, without ever having to use a third one. Latour seems to think that this third frame is of crucial importance from a physical point of view, since he writes, in an endnote:

---

[154]Let us note in passing that Latour copies these equations incorrectly (p. 18, Figure 8). It should be v/c^2 instead of v^2/c^2 in the numerator of the last equation.

> Most of the difficulties related to the ancient history of the in-
> ertia principle are related to the existence of two frames only;
> the solution is always to add a third frame that collects the in-
> formation sent by the two others. (p. 43)

Not only does Einstein never mention a third frame of reference, but in Galilean or Newtonian mechanics, to which Latour
is probably alluding when he mentions "the ancient history of
the inertia principle", this third frame does not appear either.[155]

In the same spirit, Latour lays great stress on the role of
human observers, which he analyzes in sociological terms,
evoking Einstein's purported

> obsession with transporting *in*formation through *trans*for-
> mations without *de*formation; his passion for the precise su-
> perimposition of readings; his panic at the idea that observers
> sent away might betray, might retain privileges, and send re-
> ports that could not be used to expand our knowledge; his de-
> sire to discipline the delegated observers and to turn them
> into dependent pieces of apparatus that do nothing but watch
> the coincidence of hands and notches . . . (p. 22, italics in the
> original)

But, for Einstein, the "observers" are a pedagogical fiction and
can perfectly well be replaced by apparatus; there is absolutely
no need to "discipline" them. Latour also writes:

> The ability of the delegated observers to send superimpos-
> able reports is made possible by their utter dependence and
> even stupidity. The only thing required of them is to watch
> the hands of their clocks closely and obstinately . . . That is the

---

[155]Mermin (1997b) points out, correctly, that certain technical arguments in the theory
of relativity involve comparing three (or more) frames of reference. But this has
nothing to do with Latour's purported "third frame that collects the information sent
by the two others".

price to pay for the freedom and credibility of the enunciator.
(p. 19)

In the foregoing passages, as well as in the remainder of his
paper, Latour makes a third mistake: he emphasizes the alleged
role of the "enunciator" (author) in relativity theory. But this
idea is based on a fundamental confusion between Einstein's
pedagogy and the theory of relativity itself. Einstein describes
how the space-time coordinates of an event may be transformed
from any reference frame to any other by means of the Lorentz
transformations. No reference frame plays any privileged role
here; nor does the author (Einstein) exist at all—much less
constitute a "reference frame"—*within* the physical situation he
is describing. In a certain sense, the sociological bias of Latour
has led him to misunderstand one of the fundamental tenets of
relativity, namely that no inertial reference frame is privileged
over any other.

Finally, Latour draws an eminently sensible distinction between "relativism" and "relativity": in the former, points of view
are subjective and irreconcilable; in the latter, space-time coordinates can be transformed unambiguously between reference
frames (pp. 13–14). But he then claims that the "enunciator"
plays a central role in relativity theory, which he renders in sociological and even economic terms:

[I]t is only when the enunciator's *gain* is taken into account
that the difference between relativism and relativity reveals
its deeper meaning. . . . It is the enunciator that has the privilege of accumulating all the descriptions of all the scenes he
has delegated observers to. The above dilemma boils down to
a struggle for the control of privileges, for the disciplining of
docile bodies, as Foucault would have said. (p. 15, italics in
the original)

And even more starkly:

[T]hese fights against privileges in economics or in physics are literally, and not metaphorically, the same.[156] . . . Who is going to benefit from sending all these delegated observers to the embankment, trains, rays of light, sun, nearby stars, accelerated lifts, the confines of cosmos? If relativism is right, each one of them will benefit as much as any other. If relativity is right, only *one* of them (that is, the enunciator, Einstein or some other physicist) will be able to accumulate in one place (his laboratory, his office) the documents, reports and measurements sent back by all his delegates. (p. 23, italics in the original)

This last error is rather important, since the sociological conclusions Latour wants to draw from his analysis of the theory of relativity are based on the privileged role he attributes to the "enunciator", which is in turn related to his notion of "centres of calculation".[157]

In conclusion, Latour confuses the pedagogy of relativity with the "technical content" of the theory itself. His analysis of Einstein's semi-popular book could, at best, elucidate Einstein's pedagogical and rhetorical strategies—an interesting project, to be sure, albeit a rather more modest one than showing that relativity theory is *itself* "social through and through". But, even to analyze the pedagogy fruitfully, one needs to understand the underlying theory in order to disentangle the rhetorical strategies from the physics content in Einstein's text. Latour's analysis is fatally flawed by his inadequate understanding of the theory Einstein is trying to explain.

Let us note that Latour rejects contemptuously the comments of scientists about his work:

First, the opinions of scientists about science studies are not of much importance. Scientists are the informants for our in-

---

[156]Let us note that Latour, like Lacan (see p. 20), insists here on the *literal* validity of a comparison that could, at best, be taken as a vague metaphor.

[157]This notion arises in Latour's sociology.

vestigations of science, not our judges. The vision we develop
of science does not have to resemble what scientists think
about science . . . (Latour 1995, p.6)

One may agree with the last statement. But what should one
think of an "investigator" who misunderstands so badly what his
"informants" tell him?

Latour concludes his analysis of the theory of relativity by
asking modestly:

> Did we teach Einstein anything? . . . My claim would be that,
> without the enunciator's position (hidden in Einstein's ac-
> count), and without the notion of centres of calculation, Ein-
> stein's own technical argument is ununderstandable . . . (La-
> tour 1988, p. 35)

Postscript

Almost simultaneously with the publication of our book in
France, the American journal *Physics Today* carried an essay by
physicist N. David Mermin proposing a sympathetic reading of
Latour's article on relativity and taking issue, at least implicitly,
with our own rather more critical analysis.[158] Basically, Mermin
says that criticisms of Latour's misunderstandings of relativity
miss the point, which, according to his "uniquely qualified
daughter Liz, who has been in cultural studies for some years",
is as follows:

> Latour wants to suggest translating the formal properties of
> Einstein's argument into social science, in order to see both
> what social scientists can learn about "society" and how they
> use the term, and what hard scientists can learn about their
> own assumptions. He is trying to explain relativity only inso-

---

[158]Mermin (1997b).

far as he wants to come up with a formal ("semiotic") reading
of it that can be transferred to society. He's looking for a
model for understanding social reality that will help social
scientists deal with their debates—which have to do with the
position and significance of the observer, with the relation
between "content" of a social activity and "context" (to use his
terms), and with the kinds of conclusions and rules that can
be extracted through observation. (Mermin 1997b, p. 13)

This is half-true. Latour, in his introduction, sets forth *two* goals:

[O]ur purpose . . . is the following: in what ways can we, by re-
formulating the concept of society, see Einstein's work as *explicitly* social? A related question is: how can we learn from
Einstein how to study society? (Latour 1988, p. 5, italics in
the original; see pp. 35–36 for similar statements)

For brevity, we have refrained from analyzing the extent to
which Latour achieves either of these goals, and have confined
ourselves to pointing out the fundamental misconceptions
about relativity that undermine *both* of his projects. But since
Mermin has raised the question, let us address it: Has Latour
learned anything from his analysis of relativity that can be
"transferred to society"?

At a purely logical level, the answer is no: relativity theory
in physics has no implications whatsoever for sociology. (Suppose that tomorrow an experiment at CERN were to demonstrate that the relation between an electron's velocity and its
energy is slightly different from that predicted by Einstein. This
finding would cause a revolution in physics; but why on earth
should it oblige sociologists to alter their theories of human behavior?) Clearly, the connection between relativity and sociology is, at best, one of analogy. Perhaps, by understanding the
roles of "observers" and "frames of reference" in relativity theory, Latour can shed light on sociological relativism and related
issues. But the question is who is speaking and to whom. Let's
assume, for the sake of the argument, that the sociological no-

tions used by Latour can be defined as precisely as the concepts of relativity theory and that someone familiar with both theories can establish some formal analogy between the two. This analogy might help in explaining relativity theory to a sociologist familiar with Latour's sociology, or in explaining his sociology to a physicist, but what is the point of using the analogy with relativity to explain Latour's sociology *to other sociologists*? After all, even granting Latour a complete mastery of the theory of relativity[159], his sociologist colleagues cannot be presumed to possess such knowledge. Typically, their understanding of relativity (unless they happen to have studied physics) will be based on analogies with sociological concepts. Why doesn't Latour explain whatever new sociological notions he wants to introduce by making direct reference to his readers' sociological background?

---

[159]Mermin doesn't go that far: he concedes that "there are, to be sure, many obscure statements that appear to be about the physics of relativity, which may well be misconstruals of elementary technical points." (Mermin 1997b, p. 13)

7. Intermezzo: Chaos Theory and "Postmodern Science"

> The day will come when, by study pursued through several ages,
> the things now concealed will appear with evidence; and posterity
> will be astonished that truths so clear had escaped us.
>
> *—Seneca on the motion of comets,*
> *cited by Laplace (1902 [1825], p. 6)*

One encounters frequently, in postmodernist writings, the claim that more-or-less recent scientific developments have not only modified our view of the world but have also brought about profound philosophical and epistemological shifts—in short, that the very nature of science has changed.[160] The examples cited most frequently in support of this thesis are quantum mechanics, Gödel's theorem, and chaos theory. But one also finds the arrow of time, self-organization, fractal geometry, the Big Bang, and assorted other theories.

We think that these ideas are based mostly on confusions, which are, however, much subtler than those of Lacan, Irigaray, or Deleuze. Several books would be needed to disentangle all the misunderstandings and to do justice to the kernels of truth which sometimes lie at their core. In this chapter we shall sketch such a critique, limiting ourselves to two examples: "postmodern science" according to Lyotard, and chaos theory.[161]

A by-now-classic formulation of the idea of a profound conceptual revolution can be found in Jean-François Lyotard's *The Postmodern Condition*, in a chapter devoted to "postmodern

---

[160]Numerous examples of such texts are cited in Sokal's parody (see Appendix A).

[161]See also Bricmont (1995a) for a detailed study of confusions concerning the "arrow of time".

science as the search for instabilities".[162] Here Lyotard examines some aspects of twentieth-century science that indicate, in his view, a transition towards a new "postmodern" science. Let us examine some of the examples he puts forward to support this interpretation.

After a fleeting allusion to Gödel's theorem, Lyotard addresses the limits of predictability in atomic and quantum physics. On the one hand, he observes that it is impossible, in practice, to know the positions of all the molecules in a gas, as there are vastly too many of them.[163] But this fact is well known, and it has served as the basis for statistical physics since at least the last decades of the nineteenth century. On the other hand, while apparently discussing indeterminism in quantum mechanics, Lyotard uses as an illustration a perfectly classical (non-quantum) example: the density of a gas (the ratio mass/volume). Quoting from a passage in French physicist Jean Perrin's semi-popular book on atomic physics[164], Lyotard observes that the density of a gas depends upon the scale at which the gas is observed: for example, if one considers a region whose size is comparable to that of a molecule, the density within that region may vary from zero to a very high value, depending on whether a molecule happens to be in that region or not. But this observation is banal: the density, being a macroscopic quantity, is meaningful only when a large number of molecules are involved. Nevertheless, the conceptual conclusions Lyotard draws are rather radical:

> Knowledge about the density of air thus resolves into a multiplicity of absolutely incompatible statements; they can only be made compatible if they are relativized in relation to a scale chosen by the speaker. (Lyotard 1984, p. 57)

---

[162]Lyotard (1984, chapter 13).

[163]In each cubic centimeter of air, there are approximately 2.7×10^{19} (= 27 billion billion) molecules.

[164]Perrin (1990 [1913], pp. xii–xiv).

There is a subjectivist tone in this remark, which is not justified by the case at hand. Clearly, the truth or falsity of any statement depends on the meaning of the words used. And when the meaning of those words (like density) depends on the scale, the truth or falsity of the statement will depend on the scale as well. The "multiplicity of statements" on the density of the air, when expressed carefully (i.e., specifying clearly the scale to which the statement refers), are perfectly compatible.

Later in the chapter, Lyotard mentions *fractal geometry*, which deals with "irregular" objects such as snowflakes and coastlines. These objects have, in a certain technical sense, a geometrical dimension that is not a whole number.[165] In a similar vein, Lyotard invokes *catastrophe theory*, a branch of mathematics which is devoted, roughly speaking, to classifying the cusps of certain surfaces (and similar objects). These two mathematical theories are certainly interesting, and they have had some applications in the natural sciences, notably in physics.[166] Like all scientific advances, they have provided new tools and focussed attention on new problems. But they have in no way called into question traditional scientific epistemology.

The bottom line is that Lyotard provides no argument to support his philosophical conclusions:

> The conclusion we can draw from this research (and much more not mentioned here) is that the continuous differentiable function[167] is losing its preeminence as a paradigm of

---

[165]Ordinary (smooth) geometric objects can be classified according to their *dimension*, which is always a whole number: for example, the dimension of a straight line or a smooth curve is equal to 1, while that of a plane or a smooth surface is equal to 2. By contrast, fractal objects are more complicated, and need to be assigned several distinct "dimensions" to describe different aspects of their geometry. Thus, while the "topological dimension" of any geometrical object (smooth or not) is always a whole number, the "Hausdorff dimension" of a fractal object is in general *not* a whole number.

[166]However, some physicists and mathematicians believe that the media hype surrounding these two theories has vastly exceeded their scientific accomplishments: see, for example, Zahler and Sussmann (1977), Sussmann and Zahler (1978), Kadanoff (1986) and Arnol'd (1992).

[167]These are technical concepts from differential calculus: a function is called *continuous* if (here we are oversimplifying a bit) its graph can be drawn without

knowledge and prediction. Postmodern science—by concerning itself with such things as undecidables, the limits of precise control, conflicts characterized by incomplete information, "*fracta*," catastrophes, and pragmatic paradoxes—is theorizing its own evolution as discontinuous, catastrophic, nonrectifiable[168], and paradoxical. It is changing the meaning of the word *knowledge*, while expressing how such a change can take place. It is producing not the known, but the unknown. And it suggests a model of legitimation that has nothing to do with maximized performance, but has as its basis difference understood as paralogy. (Lyotard 1984, p. 60)

Since this paragraph is frequently quoted, let us examine it closely.[169] Lyotard has here thrown together at least six distinct branches of mathematics and physics, which are conceptually quite distant from one another. Moreover, he has confused the introduction of nondifferentiable (or even discontinuous) functions in scientific models with a so-called "discontinuous" or "paradoxical" evolution of science itself. The theories cited by Lyotard of course produce new knowledge, but they do so without changing the meaning of the word.[170] *A fortiori*, what they produce is known, not unknown (except in the trivial sense that new discoveries open up new problems). Finally, the "model of legitimation" remains the comparison of theories with observa-

---

taking the pencil off the paper, while a function is called *differentiable* if, at each point of its graph, there exists a unique tangent line. Let us note in passing that every differentiable function is automatically continuous, and that catastrophe theory is based on very beautiful mathematics concerning (ironically for Lyotard) differentiable functions.

[168]"Non-rectifiable" is another technical term from differential calculus; it applies to certain non-differentiable curves.

[169]See also Bouveresse (1984, pp. 125–130) for a critique along similar lines.

[170]With one small qualification: Metatheorems in mathematical logic, such as Gödel's theorem or independence theorems in set theory, have a logical status that is slightly different from that of conventional mathematical theorems. It should, however, be emphasized that these rarefied branches of the foundations of mathematics have very little impact on the bulk of mathematical research and almost no impact on the natural sciences.

tions and experiments, not "difference understood as paralogy" (whatever this may mean).

Let us now turn our attention to chaos theory.[171] We shall address three sorts of confusions: those concerning the theory's philosophical implications, those arising from the metaphorical use of the words "linear" and "nonlinear", and those associated with hasty applications and extrapolations.

What is chaos theory about? There are many physical phenomena governed[172] by deterministic laws, and therefore predictable in principle, which are nevertheless unpredictable in practice because of their "sensitivity to initial conditions". This means that two systems obeying the same laws may, at some moment in time, be in very similar (but not identical) states and yet, after a brief lapse of time, find themselves in very different states. This phenomenon is expressed figuratively by saying that a butterfly flapping its wings today in Madagascar could provoke a hurricane three weeks from now in Florida. Of course, the butterfly by itself doesn't do much. But if one compares the two systems constituted by the Earth's atmosphere with and without the flap of the butterfly's wings, the result three weeks from now may be very different (a hurricane or not). One practical consequence of this is that we do not expect to be able to predict the weather more than a few weeks ahead.[173] Indeed, one would have to take into account such a vast quantity of data, and with such a precision, that even the largest conceivable computers could not begin to cope.

To be more precise, let us consider a system whose initial state is imperfectly known (as is always the case in practice). It

---

[171]For a deeper but still non-technical discussion, see Ruelle (1991).

[172]At least to a very high degree of approximation.

[173]Note that this does not rule out, *a priori*, the possibility of *statistically* predicting the future climate, such as the average and fluctuations in temperature and rainfall for England during the decade 2050–2060. Modelling the global climate is a difficult and controversial scientific problem, but is extremely important for the future of the human race.

is obvious that this imprecision in the initial data will be reflected in the quality of the predictions we are able to make about the system's future state. In general, the predictions will become more inexact as time goes on. But the *manner* in which the imprecision increases differs from one system to another: in some systems it will increase slowly, in others very quickly.[174]

To explain this, let us imagine that we want to reach a certain specified precision in our final predictions, and let us ask ourselves how long our predictions will remain sufficiently accurate. Let us suppose, moreover, that a technical improvement has allowed us to reduce by half the imprecision of our knowledge of the initial state. For the first type of system (where the imprecision increases slowly), the technical improvement will permit us to *double* the length of time during which we can predict the state of the system with the desired precision. But for the second type of system (where the imprecision increases quickly), it will allow us to increase our "window of predictability" by only a fixed amount: for example, by one additional hour or one additional week (how much depends on the circumstances). Simplifying somewhat, we shall call systems of the first kind *non-chaotic* and systems of the second kind *chaotic* (or "sensitive to initial conditions"). Chaotic systems are therefore characterized by the fact that their predictability is sharply limited, because even a spectacular improvement in the precision of the initial data (for example, by a factor of 1000) leads only to a rather mediocre increase in the duration over which the predictions remain valid.[175]

It is perhaps not surprising that a very complex system, such as the Earth's atmosphere, is difficult to predict. What is more surprising is that a system describable by a *small* number of

---

[174]In technical terms: in the first case the imprecision increases linearly or polynomially with time, and in the second case exponentially.

[175]It is important to add one qualification: for some chaotic systems, the fixed amount that one gains when doubling the precision in the initial measurements can be very long, which means that in practice these systems can be predictable much longer than most non-chaotic systems. For example, recent research has shown that the orbits of some planets have a chaotic behavior, but the "fixed amount" is here of the order of several million years.

variables and obeying simple deterministic equations—for example, a pair of pendulums attached together—may nevertheless exhibit very complex behavior and an extreme sensitivity to initial conditions.

However, one should avoid jumping to hasty philosophical conclusions.[176] For example, it is frequently asserted that chaos theory has shown the limits of science. But many systems in Nature are non-chaotic; and even when studying chaotic systems, scientists do not find themselves at a dead end, or at a barrier which says "forbidden to go further". Chaos theory opens up a vast area for future research and draws attention to many new objects of study.[177] Besides, thoughtful scientists have always known that they cannot hope to predict or compute *every-thing*. It is perhaps unpleasant to learn that a specific object of interest (such as the weather three weeks hence) escapes our ability to predict it, but this does not halt the development of science. For example, physicists in the nineteenth century knew perfectly well that it is impossible in practice to know the positions of all the molecules of a gas. This spurred them to develop the methods of statistical physics, which have led to an understanding of many properties of systems (such as gases) that are composed of a large number of molecules. Similar statistical methods are employed nowadays to study chaotic phenomena. And, most importantly, the aim of science is not only to predict, but also to understand.

A second confusion concerns Laplace and determinism. Let us emphasize that in this long-standing debate it has always been essential to distinguish between determinism and predictability. Determinism depends on what Nature does (independently of us), while predictability depends in part on Nature and in part on us. To see this, let us imagine a perfectly predictable phenomenon—a clock, for example—which is, however, situated in an inaccessible place (say, the top of a

---

[176]Kellert (1993) gives a clear introduction to chaos theory and a sober examination of its philosophical implications, although we do not agree with all of his conclusions.

[177]Strange attractors, Lyapunov exponents, etc.

mountain). The motion of the clock is unpredictable, *for us*, because we have no way of knowing its initial state. But it would be ridiculous to say that the clock's motion ceases to be deterministic. Or to take another example, consider a pendulum: When there is no external force, its motion is deterministic and non-chaotic. When one applies a periodic force, its motion may become chaotic and thus much more difficult to predict; but does it cease to be deterministic?

Laplace's work is often misunderstood. When he introduced the concept of universal determinism[178], he immediately added that *we* shall "always remain infinitely removed" from this imaginary "intelligence" and its ideal knowledge of the "respective situation of the beings who compose" the natural world, that is, in modern language, of the precise initial conditions of all the particles. He distinguished clearly between what Nature does and the knowledge we have of it. Moreover, he stated this principle at the beginning of an essay on *probability theory*. But, what is probability theory for Laplace? Nothing but a method that allows us to reason in situations of partial ignorance. The meaning of Laplace's text is completely misrepresented if one imagines that *he* hoped to arrive someday at a perfect knowledge and a universal predictability, for the aim of his essay was precisely to explain how to proceed in the absence of such a perfect knowledge—as one does, for example, in statistical physics.

Over the past three decades, remarkable progress has been made in the mathematical theory of chaos, but the idea that some physical systems may exhibit a sensitivity to initial conditions is not new. Here is what James Clerk Maxwell said in 1877, after stating the principle of determinism ("the same causes will always produce the same effects"):

---

[178]"Given for one instant an intelligence which could comprehend all the forces by which nature is animated and the respective situation of the beings who compose it—an intelligence sufficiently vast to submit these data to analysis—it would embrace in the same formula the movements of the greatest bodies of the universe and those of the lightest atom; for it, nothing would be uncertain and the future, as the past, would be present to its eyes." (Laplace 1902 [1825], p. 4)

There is another maxim which must not be confounded
with [this one], which asserts "That like causes produce like
effects."

This is only true when small variations in the initial circumstances produce only small variations in the final state of
the system. In a great many physical phenomena this condition is satisfied; but there are other cases in which a small
initial variation may produce a very great change in the final
state of the system, as when the displacement of the "points"
causes a railway train to run into another instead of keeping
its proper course. (Maxwell 1952 [1877], pp. 13–14)[179]

And, with regard to meteorological predictions, Henri Poincaré
in 1909 was remarkably modern:

Why have meteorologists such difficulty in predicting the
weather with any certainty? Why is it that showers and even
storms seem to come by chance, so that many people think it
quite natural to pray for rain or fine weather, though they
would consider it ridiculous to ask for an eclipse by prayer?
We see that great disturbances are generally produced in re-
gions where the atmosphere is in unstable equilibrium. The
meteorologists see very well that the equilibrium is unstable,
that a cyclone will be formed somewhere, but exactly where
they are not in a position to say; a tenth of a degree more or
less at any given point, and the cyclone will burst here and not
there, and extend its ravages over districts it would otherwise
have spared. If they had been aware of this tenth of a degree,
they could have known it beforehand, but the observations
were neither sufficiently comprehensive nor sufficiently precise, and that is the reason why it all seems due to the intervention of chance. (Poincaré 1914 [1909], pp. 68–69)

---

[179] The purpose of quoting these remarks is, of course, to clarify the distinction
between determinism and predictability, not to prove that determinism is true.
Indeed, Maxwell himself was apparently not a determinist.

Let us turn now to the confusions arising from misuse of the words "linear" and "nonlinear". Let us first point out that, in mathematics, the word "linear" has two distinct meanings, which it is important not to confuse. On the one hand, one may speak of a *linear function* (or *equation*): for example, the functions $f(x) = 2x$ and $f(x) = -17x$ are linear, while the functions $f(x) = x^2$ and $f(x) = \sin x$ are nonlinear. In terms of mathematical modelling, a linear equation describes a situation in which (simplifying somewhat) "the effect is proportional to the cause".[180] On the other hand, one may speak of a *linear order*[181]: this means that the elements of a set are ordered in such a way that, for each pair of elements a and b, one has either $a < b$, $a = b$, or $a > b$. For instance, there exists a natural linear order on the set of real numbers, while there is no natural such order on the complex numbers.[182] Now, postmodernist authors (principally in the English-speaking world) have added a third meaning to the word—vaguely related to the second, but often confused by them with the first—in speaking of *linear thought*. No exact definition is given, but the general meaning is clear enough: it is the logical and rationalist thought of the Enlightenment and of so-called "classical" science (often accused of an extreme reductionism and numericism). In opposition to this old-fashioned way of thinking, they advocate a postmodern "nonlinear thought". The precise content of the latter is not clearly explained either, but it is, apparently, a methodology that goes beyond reason by insisting on intuition and subjective perception.[183] And it is frequently claimed that so-called postmodern

---

[180]This verbal formulation actually confuses the problem of linearity with the very different problem of causality. In a linear equation, it is the *set of all the variables* that obeys a relation of proportionality. There is no need to specify which variables represent the "effect" and which the "cause"; and indeed, in many instances (for example, in systems with feedback) such a distinction is meaningless.

[181]Often called *total order*.

[182][For the experts:] Here "natural" means "compatible with the field structure", in the sense that $a, b > 0$ implies $ab > 0$, and $a > b$ implies $a + c > b + c$.

[183]Let us note in passing that it is *false* to assert that intuition plays no role in "traditional" science. Quite the contrary: since scientific theories are creations of the human mind and are almost never "written" in the experimental data, intuition plays

science—and particularly chaos theory—justifies and supports this new "nonlinear thought". But this assertion rests simply on a confusion between the three meanings of the word "linear".[184]

Because of these abuses, one often finds postmodernist authors who see chaos theory as a revolution against Newtonian mechanics—the latter being labelled "linear"—or who cite quantum mechanics as an example of a nonlinear theory.[185] In actual fact, Newton's "linear thought" uses equations that are perfectly *nonlinear*; this is why many examples in chaos theory come from Newtonian mechanics, so that the study of chaos represents in fact a *renaissance* of Newtonian mechanics as a subject for cutting-edge research. Likewise, quantum mechanics

---

an essential role in the creative process of *invention* of theories. Nevertheless, intuition cannot play an explicit role in the reasoning leading to the *verification* (or falsification) of these theories, since this process must remain independent of the subjectivity of individual scientists.

[184]For example: "These [scientific] practices were rooted in a binary logic of hermetic subjects and objects and a linear, teleological rationality . . . Linearity and teleology are being supplanted by chaos models of non-linearity and an emphasis on historical contingency." (Lather 1991, pp. 104–105)

"As opposed to more linear (historical and psychoanalytic as well as scientific) determinisms that tend to exclude them as anomalies outside the generally linear course of things, certain older determinisms incorporated chaos, incessant turbulence, sheer chance, in dynamic interactions cognate to modern chaos theory . . ." (Hawkins 1995, p. 49)

"Unlike teleological linear systems, chaotic models resist closure, breaking off instead into endless 'recursive symmetries.' This lack of closure privileges uncertainty. A single theory or 'meaning' disseminates into infinite possibilities . . . What we once considered to be enclosed by linear logic begins to open up to a surprising series of new forms and possibilities." (Rosenberg 1992, p. 210)

Let us emphasize that we are *not* criticizing these authors for employing the word "linear" in their own sense: mathematics has no monopoly on the word. What we are criticizing is some postmodernists' tendency to *confuse* their sense of the word with the mathematical one, and to draw connections with chaos theory that are not supported by any valid argument. Dahan-Dalmedico (1997) seems to miss this point.

[185]For example, Harriett Hawkins refers to the "linear equations describing the regular, and therefore predictable movements of planets and comets" (Hawkins 1995, p. 31), and Steven Best alludes to "the linear equations used in Newtonian and even quantum mechanics" (Best 1991, p. 225); they commit the first mistake but not the second. Conversely, Robert Markley claims that "Quantum physics, hadron bootstrap theory, complex number theory [!], and chaos theory share the basic assumption that reality cannot be described in linear terms, that nonlinear—and unsolvable— equations are the only means possible to describe a complex, chaotic, and non-deterministic reality." (Markley 1992, p. 264) This sentence deserves some sort of prize for squeezing the maximal number of confusions into the minimal number of words. See p. 266 below for a brief discussion.

is often cited as the quintessential example of a "postmodern science", but the fundamental equation of quantum mechanics—Schrödinger's equation—is absolutely *linear*.

Furthermore, the relationship between linearity, chaos, and an equation's explicit solvability is often misunderstood. Nonlinear equations are generally more difficult to solve than linear equations, but not always: there exist very difficult linear problems and very simple nonlinear ones. For example, Newton's equations for the two-body Kepler problem (the Sun and *one* planet) are nonlinear and yet explicitly solvable. Besides, for chaos to occur, it is necessary that the equation be nonlinear and (here we simplify somewhat) not explicitly solvable, but these two conditions are by no means *sufficient*—whether they occur separately or together—to produce chaos. Contrary to what people often think, a nonlinear system is not necessarily chaotic.

The difficulties and confusions multiply when one attempts to apply the mathematical theory of chaos to concrete situations in physics, biology, or the social sciences.[186] To do this in a sensible way, one must first have some idea of the relevant variables and of the type of evolution they obey. Unfortunately, it is often difficult to find a mathematical model that is sufficiently simple to be analyzable and yet adequately describes the objects being considered. These problems arise, in fact, whenever one tries to apply a mathematical theory to reality.

Some purported "applications" of chaos theory—for example, to business management or literary analysis—border on the absurd.[187] And, to make things worse, chaos theory—which is well-developed mathematically—is often confused with the still-emerging theories of complexity and self-organization.

Another major confusion is caused by mixing the mathematical theory of chaos with the popular wisdom that small causes can have large effects: "if Cleopatra's nose had been

---

[186]See Ruelle (1994) for a more detailed discussion.

[187]For thoughtful critiques of applications of chaos theory in literature, see, for example, Matheson and Kirchhoff (1997) and van Peer (1998).

shorter", or the story of the missing nail that led to the collapse of an empire. One constantly hears claims of chaos theory being "applied" to history or society. But human societies are complicated systems involving a vast number of variables, for which one is unable (at least at present) to write down any sensible equations. To speak of chaos for these systems does not take us much further than the intuition already contained in the popular wisdom.[188]

Yet another abuse arises from confusing (intentionally or not) the numerous distinct meanings of the highly evocative word "chaos": its technical meaning in the mathematical theory of nonlinear dynamics—where it is roughly (though not exactly) synonymous with "sensitive dependence on initial conditions"—and its wider senses in sociology, politics, history and theology, where it is frequently taken as a synonym for disorder. As we shall see, Baudrillard and Deleuze–Guattari are especially shameless in exploiting (or falling into) these verbal confusions.

---

[188]We do not deny that if one understood these systems better—enough to be able to write down equations that describe them at least approximately—the mathematical theory of chaos might provide interesting information. But sociology and history are, at present, far from having reached this stage of development (and perhaps will always remain so).

8. Jean Baudrillard

Jean Baudrillard's sociological work challenges and provokes all
current theories. With derision, but also with *extreme precision*,
he unknots the constituted social descriptions with quiet
confidence and a sense of humor.
                    —*Le Monde (1984b, p. 95, italics added)*

The sociologist and philosopher Jean Baudrillard is well-known
for his reflections on the problems of reality, appearance, and illusion. In this chapter we want to draw attention to a less-noted
aspect of Baudrillard's work, namely his frequent use of scientific and pseudo-scientific terminology.

In some cases, Baudrillard's invocation of scientific concepts is clearly metaphorical. For example, he wrote about the
Gulf War as follows:

> What is most extraordinary is that the two hypotheses,
> the apocalypse of real time and pure war along with the tri-
> umph of the virtual over the real, are realised at the same
> time, in the same space-time, each in implacable pursuit of the
> other. It is a sign that the space of the event has become a hy-
> perspace with multiple refractivity, and that *the space of war
> has become definitively non-Euclidean.* (Baudrillard 1995,
> p. 50, italics in the original)

There seems to be a tradition of using technical mathematical
notions out of context. With Lacan, it was tori and imaginary
numbers; with Kristeva, infinite sets; and here we have non-
Euclidean spaces.[189] But what could this metaphor mean? In-

---

[189]What is a non-Euclidean space? In Euclidean plane geometry—the geometry
studied in high school—for each straight line L and each point p not on L, there

deed, what would a *Euclidean* space of war look like? Let us
note in passing that the concept of "hyperspace with multiple re-
fractivity" [*hyperespace à réfraction multiple*] does not exist in
either mathematics or physics; it is a Baudrillardian invention.

Baudrillard's writings are full of similar metaphors drawn
from mathematics and physics, for example:

> In the Euclidean space of history, the shortest path be-
> tween two points is the straight line, the line of Progress and
> Democracy. But this is only true of the linear space of the En-
> lightenment.[190] In our non-Euclidean *fin de siècle* space, a
> baleful curvature unfailingly deflects all trajectories. This is
> doubtless linked to the sphericity of time (visible on the hori-
> zon of the end of the century, just as the earth's sphericity is
> visible on the horizon at the end of the day) or the subtle dis-
> tortion of the gravitational field. . . .
>
> By this retroversion of history to infinity, this hyperbolic
> curvature, the century itself is escaping its end. (Baudrillard
> 1994, pp. 10–11)

> It is to this perhaps that we owe this 'fun physics' effect:
> the impression that events, collective or individual, have been
> bundled into a memory hole. This blackout is due, no doubt,
> to this movement of reversal, this parabolic curvature of his-
> torical space. (Baudrillard 1994, p. 20)

But not all of Baudrillard's physics is metaphorical. In his
more philosophical texts, Baudrillard apparently takes
physics—or his version of it—literally, as in his essay "The
fatal, or, reversible imminence", devoted to the theme of chance:

---

exists one and only one straight line parallel to L (i.e., not intersecting L) that passes
through p. By contrast, in non-Euclidean geometries, there can be either an infinite
number of parallel lines or else none at all. These geometries go back to the works of
Bolyai, Lobachevskii, and Riemann in the nineteenth century, and they were applied
by Einstein in his general theory of relativity (1915). For a good introduction to non-
Euclidean geometries (but without their military applications), see Greenberg (1980)
or Davis (1993).

[190]See our discussion (p. 143–45 above) concerning abuses of the word "linear".

This reversibility of causal order—the reversion of cause on effect, the precession and triumph of effect over cause—is fundamental. . . .

This is what science catches a glimpse of when, not happy with calling into question the determinist principle of causality (the first revolution), it intuits—beyond even the uncertainty principle, which still functions like hyper-rationality—that chance is the floating of all laws. This is already quite extraordinary. But what science senses now, at the physical and biological limits of its exercise, is that there is not only this floating, this uncertainty, but a possible *reversibility* of physical laws. That would be the *absolute enigma*, not some ultra-formula or meta-equation of the universe (which the theory of relativity was), but the idea that any law can be reversed (not only particles into anti-particles, matter into anti-matter, but the laws themselves). The hypothesis of this reversibility has always been affirmed by the great metaphysical systems. It is the fundamental rule of the game of appearance, of the metamorphosis of appearances, against the irreversible order of time, of law and meaning. But it's fascinating to see science arrive at the same hypotheses, contrary to its own logic and evolution. (Baudrillard 1990, pp. 162–163, italics in the original)

It is difficult to know what Baudrillard means by "reversing" a law of physics. In physics one can speak of the laws' *reversibility*, as a shorthand for their "invariance with respect to time inversion".[191] But this property is already well-known in Newtonian mechanics, which is as causal and deterministic as a theory can be; it has nothing to do with uncertainty and is in no way at the "physical and biological limits" of science. (Quite

---

[191]To illustrate this concept, consider a collection of billiard balls moving on a table according to Newton's laws (without friction and with elastic collisions), and make a film of this motion. Now run this film backwards: the reversed motion will also obey the laws of Newtonian mechanics. This fact is summarized by saying that the laws of Newtonian mechanics are invariant with respect to time inversion. In fact, all the known laws of physics, except those of the "weak interactions" between subatomic particles, satisfy this property of invariance.

the opposite: it is the *non*-reversibility of the laws of the "weak interactions", discovered in 1964, that is new and at present imperfectly understood.) In any case, the reversibility of the laws has nothing to do with an alleged "reversibility of causal order". Finally, Baudrillard's scientific confusions (or fantasies) have led him to make unwarranted philosophical claims: he puts forward no argument whatsoever to support his idea that science arrives at hypotheses "contrary to its own logic".

This train of thought is taken up once again in his essay entitled "Exponential instability, exponential stability":

> The whole problem of speaking about the end (particularly the end of history) is that you have to speak of what lies beyond the end and also, at the same time, of the impossibility of ending. This paradox is produced by the fact that in a non-linear, non-Euclidean space of history the end cannot be located. The end is, in fact, only conceivable in a logical order of causality and continuity. Now, it is events themselves which, by their artificial production, their programmed occurrence or the anticipation of their effects—not to mention their transfiguration in the media—are suppressing the cause-effect relation and hence all historical continuity.
>
> This distortion of causes and effects, this mysterious autonomy of effects, this cause-effect reversibility, engendering a disorder or chaotic order (precisely our current situation: a reversibility of reality [*le réel*] and information, which gives rise to disorder in the realm of events and an extravagance of media effects), puts one in mind, to some extent, of Chaos Theory and the disproportion between the beating of a butterfly's wings and the hurricane this unleashes on the other side of the world. It also calls to mind Jacques Benveniste's paradoxical hypothesis of the memory of water. . . .
>
> Perhaps history itself has to be regarded as a chaotic formation, in which acceleration puts an end to linearity and the turbulence created by acceleration deflects history definitively from its end, just as such turbulence distances effects from their causes. (Baudrillard 1994, pp. 110–111)

First of all, chaos theory in no way reverses the relationship between cause and effect. (Even in human affairs, we seriously doubt that an action in the present could affect an event in the *past!*) Moreover, chaos theory has nothing to do with Benveniste's hypothesis on the memory of water.[192] And finally, the last sentence, though constructed from scientific terminology, is meaningless from a scientific point of view.

The text continues in a gradual crescendo of nonsense:

We shall not reach the destination, even if that destination is the Last Judgment, since we are henceforth separated from it by a variable refraction hyperspace. The retroversion of history could very well be interpreted as a turbulence of this kind, due to the hastening of events which reverses and swallows up their course. This is one of the versions of Chaos Theory—that of *exponential instability* and its uncontrollable effects. It accounts very well for the 'end' of history, interrupted in its linear or dialectical movement by that catastrophic singularity . . .

But the exponential instability version is not the only one. The other is that of *exponential stability*. This latter defines a state in which, no matter where you start out, you always end up at the same point. The initial conditions, the original singularities do not matter: everything tends towards the Zero point—itself also a strange attractor.[193] . . .

Though incompatible, the two hypotheses—exponential instability and stability—are in fact simultaneously valid.

[192]The experiments of Benveniste's group on the biological effects of highly diluted solutions, which seemed to provide a scientific basis for homeopathy, were rapidly discredited after being hastily published in the scientific journal *Nature* (Davenas *et al.* 1988). See Maddox *et al.* (1988); and, for a more detailed discussion, see Broch (1992). More recently, Baudrillard has opined that the memory of water is "the ultimate stage of the transfiguration of the world into pure information" and that "this virtualization of effects is wholly in line with the most recent science." (Baudrillard 1997, p. 94)

[193]Not at all! When zero is an attractor, it is what one calls a "fixed point"; these attractors (as well as others known as "limit-cycles") have been known since the nineteenth century, and the expression "strange attractor" was introduced specifically to refer to attractors of a different sort. See, for example, Ruelle (1991).

Moreover, our system, in its *normal*—normally catastrophic—course combines them very well. It combines in
effect an inflation, a galloping acceleration, a dizzying whirl of
mobility, an eccentricity of events and an excess of meaning
and information with an exponential tendency towards total
entropy. Our systems are thus doubly chaotic: they operate
both by exponential stability and instability.

It would seem then that there will be no end because we
are already in an excess of ends: the transfinite. . . .

Our complex, metastatic, viral systems, condemned to
the exponential dimension alone (be it that of exponential
stability or instability), to eccentricity and indefinite fractal
scissiparity, can no longer come to an end. Condemned to an
intense metabolism, to an intense internal metastasis, they
become exhausted within themselves and no longer have any
destination, any end, any otherness, any fatality. They are
condemned, precisely, to the epidemic, to the endless ex-
crescences of the fractal and not to the reversibility and perfect resolution of the fateful [*fatal*]. We know only the signs of
catastrophe now; we no longer know the signs of destiny.
(And besides, has any concern been shown in Chaos Theory
for the equally extraordinary, contrary phenomenon of *hy-
posensitivity* to initial conditions, of the inverse exponen-
tiality of effects in relation to causes—the potential
hurricanes which end in the beating of a butterfly's wings?)
(Baudrillard 1994, pp. 111–114, italics in the original)

The last paragraph is Baudrillardian *par excellence*. One would
be hard pressed not to notice the high density of scientific and
pseudo-scientific terminology[194]—inserted in sentences that are,
as far as we can make out, devoid of meaning.

These texts are, however, atypical of Baudrillard's œuvre,
because they allude (albeit in a confused fashion) to more-or-
less well-defined scientific ideas. More often one comes across
sentences like these:

---

[194]Examples of the latter are *variable refraction hyperspace* and *fractal scissiparity*.

There is no better model of the way in which the computer screen and the mental screen of our brain are interwoven than Moebius's topology, with its peculiar contiguity of near and far, inside and outside, object and subject within the same spiral. It is in accordance with this same model that information and communication are constantly turning round upon themselves in an incestuous circumvolution, a superficial conflation of subject and object, within and without, question and answer, event and image, and so on. The form is inevitably that of a twisted ring reminiscent of the mathematical symbol for infinity. (Baudrillard 1993, p. 56)

As Gross and Levitt remark, "this is as pompous as it is meaningless."[195]

In summary, one finds in Baudrillard's works a profusion of scientific terms, used with total disregard for their meaning and, above all, in a context where they are manifestly irrelevant.[196] Whether or not one interprets them as metaphors, it is hard to see what role they could play, except to give an appearance of profundity to trite observations about sociology or history. Moreover, the scientific terminology is mixed up with a nonscientific vocabulary that is employed with equal sloppiness. When all is said and done, one wonders what would be left of Baudrillard's thought if the verbal veneer covering it were stripped away.[197]

---

[195]Gross and Levitt (1994, p. 80).

[196]For other examples, see the references to chaos theory (Baudrillard 1990, pp. 154–155), to the Big Bang (Baudrillard 1994, pp. 115–116), and to quantum mechanics (Baudrillard 1996, pp. 14, 53–55). This last book is permeated with scientific and pseudo-scientific allusions.

[197]For a more detailed critique of Baudrillard's ideas, see Norris (1992).

9. Gilles Deleuze and Félix Guattari

> I must speak here about two books that seem to me to be among the greatest of the great: *Difference and Repetition*, *The Logic of Sense*. Undoubtedly so great, in fact, that it is difficult to speak about them and few have done so. For a long time, I believe, this work will soar over our heads, in enigmatic resonance with that of Klossovski, another major and excessive sign. But some day, perhaps, the century will be Deleuzian.
> —*Michel Foucault, Theatrum Philosophicum (1970, p. 885)*

Gilles Deleuze, who died recently, is reputed to be one of the most important contemporary French thinkers. He has written twenty-odd books of philosophy, either alone or in collaboration with the psychoanalyst Félix Guattari. In this chapter we shall analyze that part of Deleuze and Guattari's œuvre where they invoke terms and concepts from physics or mathematics.

The main characteristic of the texts quoted in this chapter is their lack of clarity. Of course, defenders of Deleuze and Guattari could retort that these texts are profound and that we have failed to understand them properly. However, on closer examination, one sees that there is a great concentration of scientific terms, employed out of context and without any apparent logic, at least if one attributes to these terms their usual scientific meanings. To be sure, Deleuze and Guattari are free to use these terms in other senses: science has no monopoly on the use of words like "chaos", "limit" or "energy". But, as we shall show, their writings are crammed also with highly technical terms that are not used outside of specialized scientific discourses, and for which they provide no alternative definition.

These texts touch on a great variety of subjects: Gödel's theorem, the theory of transfinite cardinals, Riemannian geometry, quantum mechanics . . .[198] But the allusions are so brief and superficial that a reader who is not already an expert in these subjects will be unable to learn anything concrete. And a specialist reader will find their statements most often meaningless, or sometimes acceptable but banal and confused.

We are well aware that Deleuze and Guattari's subject is philosophy, not the popularization of science. But what philosophical function can be fulfilled by this avalanche of ill-digested scientific (and pseudo-scientific) jargon? In our opinion, the most plausible explanation is that these authors possess a vast but very superficial erudition, which they put on display in their writings.

Their book *What is Philosophy?* was a best-seller in France in 1991. One of its principal themes is the distinction between philosophy and science. According to Deleuze and Guattari, philosophy deals with "concepts", while science deals with "functions". Here is how they describe this contrast:

> [T]he first difference between science and philosophy is their respective attitudes toward chaos. Chaos is defined not so much by its disorder as by the infinite speed with which every form taking shape in it vanishes. It is a void that is not a nothingness but a *virtual*, containing all possible particles and drawing out all possible forms, which spring up only to disappear immediately, without consistency or reference, without consequence. Chaos is an infinite speed of birth and disappearance. (Deleuze and Guattari 1994, pp. 117–118, italics in the original)

---

[198]Gödel: Deleuze and Guattari (1994, pp. 121, 137–139). Transfinite cardinals: Deleuze and Guattari (1994, pp. 120–121). Riemannian geometry: Deleuze and Guattari (1987, pp. 32, 373, 482–486, 556n). Deleuze and Guattari (1994, pp. 124, 161, 217). Quantum mechanics: Deleuze and Guattari (1994, pp. 129–130). These references are far from being exhaustive.

Let us note in passing that the word "chaos" is not being used here in its usual scientific sense (see Chapter 7 above)[199], although, later in the book, it is employed without comment also in this latter sense.[200] They continue as follows:

> Now philosophy wants to know how to retain infinite speeds while gaining consistency, by *giving the virtual a consistency specific to it*. The philosophical sieve, as plane of immanence that cuts through the chaos, selects infinite movements of thought and is filled with concepts formed like consistent particles going as fast as thought. Science approaches chaos in a completely different, almost opposite way: it relinquishes the infinite, infinite speed, in order to gain *a reference able to actualize the virtual*. By retaining the infinite, philosophy gives consistency to the virtual through concepts; by relinquishing the infinite, science gives a reference to the virtual, which actualizes it through functions. Philosophy proceeds with a plane of immanence or consistency; science with a plane of reference. In the case of science it is like a freeze-frame. It is

---

[199]Indeed, Deleuze and Guattari, in a footnote, refer the reader to a book by Prigogine and Stengers, where one finds the following picturesque description of quantum field theory:

> The quantum vacuum is the opposite of nothingness: far from being passive or inert, it potentially contains all possible particles. Unceasingly, these particles emerge out of the vacuum, only to disappear immediately. (Prigogine and Stengers 1988, p. 162)

A little later, Prigogine and Stengers discuss some theories on the origin of the universe that involve an instability of the quantum vacuum (in general relativity), and they add:

> This description is reminiscent of the crystallization of a supercooled liquid (a liquid that has been cooled below its freezing temperature). In such a liquid, small crystalline kernels form, but they then dissolve without consequences. For such a kernel to unleash the process leading to the crystallization of the entire liquid, it has to reach a critical size that depends, in this case too, on a highly nonlinear cooperative mechanism called "nucleation". (Prigogine and Stengers 1988, pp. 162–163)

The conception of "chaos" used by Deleuze and Guattari is thus a verbal mélange of a description of quantum field theory with a description of a supercooled liquid. These two branches of physics have no direct relation to chaos theory in its usual sense (namely, the theory of nonlinear dynamical systems).

[200]Deleuze and Guattari (1994), p. 156 and note 14, and especially p. 206 and note 7.

a fantastic *slowing down*, and it is by slowing down that matter, as well as the scientific thought able to penetrate it with
propositions, is actualized. A function is a Slow-motion. Of
course, science constantly advances accelerations, not only in
catalysis but in particle accelerators and expansions that
move galaxies apart. However, the primordial slowing down
is not for these phenomena a zero-instant with which they
break but rather a condition coextensive with their whole development. To slow down is to set a limit in chaos to which all
speeds are subject, so that they form a variable determined as
abscissa, at the same time as the limit forms a universal constant that cannot be gone beyond (for example, a maximum
degree of contraction). The first functives are therefore the
limit and the variable, and reference is a relationship between
values of the variable or, more profoundly, the relationship of
the variable, as abscissa of speeds, with the limit. (Deleuze
and Guattari 1994, pp. 118–119, italics in the original)

This passage contains at least a dozen scientific terms[201] used
without rhyme or reason, and the discourse oscillates between
nonsense ("a function is a Slow-motion") and truisms ("science
constantly advances accelerations"). But what comes next is
even more impressive:

Sometimes the constant-limit itself appears as a relationship in the whole of the universe to which all the parts are
subject under a finite condition (quantity of movement, force,
energy). Again, there must be systems of coordinates to which
the terms of the relationship refer: this, then, is a second sense
of limit, an external framing or exoreference. For these pro-
tolimits, outside all coordinates, initially generate speed ab-
scissas on which axes will be set up that can be coordinated.
A particle will have a position, an energy, a mass, and a spin
value but on condition that it receives a physical existence or

---

[201]For example: *infinite, speed, particle, function, catalysis, particle accelerator,
expansion, galaxy, limit, variable, abscissa, universal constant, contraction.*

actuality, or that it "touches down" in trajectories that can be grasped by systems of coordinates. It is these first limits that constitute slowing down in the chaos or the threshold of suspension of the infinite, which serve as endoreference and carry out a counting: they are not relations but numbers, and the entire theory of functions depends on numbers. We refer to the speed of light, absolute zero, the quantum of action, the Big Bang: the absolute zero of temperature is minus 273.15 degrees Centigrade, the speed of light, 299,796 kilometers per second, where lengths contract to zero and clocks stop. Such limits do not apply through the empirical value that they take on solely within systems of coordinates, they act primarily as the condition of primordial slowing down that, in relation to infinity, extends over the whole scale of corresponding speeds, over their conditioned accelerations or slowing-downs. It is not only the diversity of these limits that entitles us to doubt the unitary vocation of science. In fact, each limit on its own account generates irreducible, heterogeneous systems of coordinates and imposes thresholds of discontinuity depending on the proximity or distance of the variable (for example, the distance of the galaxies). Science is haunted not by its own unity but by the plane of reference constituted by all the limits or borders through which it confronts chaos. It is these borders that give the plane its references. As for the systems of coordinates, they populate or fill out the plane of reference itself. (Deleuze and Guattari 1994, pp. 119–120)

With a bit of work, one can detect in this paragraph a few meaningful phrases[202], but the discourse in which they are immersed is utterly meaningless.

The next pages are in the same genre, and we shall refrain from boring the reader with them. Let us remark, however, that not all the invocations of scientific terminology in this book are

---

[202]For example, the statement "the speed of light . . . where lengths contract to zero and clocks stop" is not false, but it may lead to confusion. In order to understand it correctly, one must already have a good knowledge of relativity theory.

quite so absurd. Some passages seem to address serious problems in the philosophy of science, for example:

> As a general rule, the observer is neither inadequate nor subjective: even in quantum physics, Heisenberg's demon does not express the impossibility of measuring both the speed and the position of a particle on the grounds of a subjective interference of the measure with the measured, but it measures exactly an objective state of affairs that leaves the respective position of two of its particles outside of the field of its actualization, the number of independent variables being reduced and the values of the coordinates having the same probability. (Deleuze and Guattari 1994, p. 129)

The beginning of this text has the aura of a deep remark on the interpretation of quantum mechanics, but the end (starting with "leaves the respective position") is totally devoid of meaning. And they continue:

> Subjectivist interpretations of thermodynamics, relativity, and quantum physics manifest the same inadequacies. Perspectivism, or scientific relativism, is never relative to a subject: it constitutes not a relativity of truth but, on the contrary, a truth of the relative, that is to say, of variables whose cases it orders according to the values it extracts from them in its system of coordinates (here the order of conic sections is ordered according to sections of the cone whose summit is occupied by the eye). (Deleuze and Guattari 1994, pp. 129–130)

Again, the end of the passage is meaningless, even if the beginning alludes vaguely to the philosophy of science.[203]

Similarly, Deleuze and Guattari appear to discuss issues in the philosophy of mathematics:

---

[203]For an amusing exegesis of the above passages, in the same vein as the original, see Alliez (1993, Chapter II).

The respective independence of variables appears in mathematics when one of them is at a higher power than the first. That is why Hegel shows that variability in the function is not confined to values that can be changed (2/3 and 4/6) or are left undetermined ($a = 2b$) but requires one of the variables to be at a higher power ($y^2/x = P$).[204] For it is then that a relation can be directly determined as differential relation dy/dx, in which the only determination of the value of the variables is that of disappearing or being born, even though it is wrested from infinite speeds. A state of affairs or "derivative" function depends on such a relation: an operation of depotentialization has been carried out that makes possible the comparison of distinct powers starting from which a thing or a body may well develop (integration). In general, a state of affairs does not actualize a chaotic virtual without taking from it a *potential* that is distributed in the system of coordinates. From the virtual that it actualizes it draws a potential that it appropriates. (Deleuze and Guattari 1994, p. 122, italics in the original)

Here Deleuze and Guattari recycle, with a few additional inventions (*infinite speeds*, *chaotic virtual*), old ideas of Deleuze's that originally appeared in the book Michel Foucault judged "among the greatest of the great", *Difference and Repetition*. At two places in this book, Deleuze discusses classical problems in the conceptual foundations of differential and integral calculus. After the birth of this branch of mathematics in the seventeenth century through the works of Newton and Leibniz, cogent objections were raised against the use of "infinitesimal" quantities such as dx and dy.[205] These problems were solved by the work of d'Alembert around 1760 and Cauchy around 1820,

---

[204]This sentence repeats a confusion of Hegel (1989 [1812], pp. 251–253, 277–278), who considered that fractions such as y^2/x were fundamentally different from fractions like a/b. As noted by the philosopher J.T. Desanti: "Such propositions could not but surprise a 'mathematical mind', who would be led to regard them as absurd." (Desanti 1975, p. 43)

[205]Which appear in the derivative dy/dx and in the integral $\int f(x)dx$.

who introduced the rigorous notion of *limit*—a concept that has been taught in all calculus textbooks since the middle of the nineteenth century.[206] Nevertheless, Deleuze launches into a long and confused meditation on these problems, from which we shall quote just a few characteristic excerpts[207]:

> Must we say that vice-diction[208] does not go as far as contradiction, on the grounds that it concerns only properties? In reality, the expression 'infinitely small difference' does indeed indicate that the difference vanishes so far as intuition is concerned. Once it finds its concept, however, it is rather intuition itself which disappears in favour of the differential relation, as is shown by saying that dx is minimal in relation to x, as dy is in relation to y, but that dy/dx is the internal qualitative relation, expressing the universal of a function independently of its particular numerical values.[209] However, if this relation has no numerical determinations, it does have degrees of variation corresponding to diverse forms and equations. These degrees are themselves like the relations of the universal, and the differential relations, in this sense, are caught up in a process of reciprocal determination which translates the interdependence of the variable coefficients. But once again, *reciprocal determination* expresses only the first aspect of a veritable principle of reason; the second aspect is *complete*

---

[206] For a historical account, see, for example, Boyer (1959 [1949], pp. 247–250, 267–277).

[207] Further comments on calculus can be found in Deleuze (1994, pp. 43, 170–178, 182–183, 201, 209–211, 244, 264, 280–281). For additional lucubrations on mathematical concepts, mixing banalities with nonsense, see Deleuze (1994, pp. 179–181, 202, 232–234, 237–238); and on physics, see Deleuze (1994, pp. 117, 222–226, 228–229, 240, 318n).

[208] The previous paragraph contains the following definition: "This procedure of the infinitely small, which maintains the distinction between essences (to the extent that one plays the role of inessential to the other), is quite different to contradiction. We should therefore give it a special name, that of 'vice-diction'." (Deleuze 1994, p. 46)

[209] This is, at best, a very complicated way of saying that the traditional notation dy/dx denotes an object—the derivative of the function $y(x)$—which is not, however, the quotient of two quantities dy and dx.

determination. For each degree or relation, regarded as the universal of a given function, determines the existence and distribution of distinctive points on the corresponding curve. We must take great care here not to confuse 'complete' with 'completed'. The difference is that, for the equation of a curve, for example, the differential relation refers only to straight lines determined by the nature of the curve. It is already a complete determination of the object, yet it expresses only a part of the entire object, namely the part regarded as 'derived' (the other part, which is expressed by the so-called primitive function, can be found only by integration, which is not simply the inverse of differentiation.[210] Similarly, it is integration which defines the nature of the previously determined distinctive points). That is why an object can be completely determined—*ens omni modo determinatum*—without, for all that, possessing the integrity which alone constitutes its actual existence. Under the double aspect of reciprocal determination and complete determination, however, it appears already as if the limit coincides with the power itself. The limit is defined by convergence. The numerical values of a function find their limit in the differential relation; the differential relations find their limit in the degrees of variation; and at each degree the distinctive points are the limits of series which are analytically continued one into the other. Not only is the differential relation the pure element of potentiality, but the limit is the power of the continuous [*puissance du continu*][211], as continuity is the power of these limits themselves. (Deleuze 1994, pp. 46–47, italics in the original)

---

[210]In the calculus of functions of a single variable, integration is indeed the inverse of differentiation, up to an additive constant (at least for sufficiently smooth functions). The situation is more complicated for functions of several variables. Conceivably Deleuze is referring to this latter case, but if so, he is doing it in a very confused fashion.

[211]The correct translation of the mathematical term *"puissance du continu"* is "power of the continuum". See note 38 above for a brief explanation of this concept.

Deleuze notwithstanding, "limit" and "power of the continuum" are two completely distinct concepts. It is true that the idea of "limit" is related to the idea of "real number", and that the *set* of the real numbers has the power of the continuum. But Deleuze's formulation is, at best, exceedingly confused.

Just as we oppose difference in itself to negativity, so we oppose dx to not-A, the symbol of difference [*Differenzphilosophie*] to that of contradiction. It is true that contradiction seeks its Idea on the side of the greatest difference, whereas the differential risks falling into the abyss of the infinitely small. This, however, is not the way to formulate the problem: it is a mistake to tie the value of the symbol dx to the existence of infinitesimals; but it is also a mistake to refuse it any ontological or gnoseological value in the name of a refusal of the latter. . . . The principle of a general differential philosophy must be the object of a rigorous exposition, and must in no way depend upon the infinitely small.[212] The symbol dx appears as simultaneously undetermined, determinable and determination. Three principles which together form a sufficient reason correspond to these three aspects: a principle of determinability corresponds to the undetermined as such (dx, dy); a principle of reciprocal determination corresponds to the really determinable (dy/dx); a principle of complete determination corresponds to the effectively determined (values of dy/dx). In short, dx is the Idea—the Platonic, Leibnizian or Kantian Idea, the 'problem' and its being. (Deleuze 1994, pp. 170–171, italics in the original)

[T]he differential relation presents a third element, that of pure potentiality. Power is the form of reciprocal determination according to which variable magnitudes are taken to be functions of one another. In consequence, calculus considers only those magnitudes where at least one is of a power superior to another.[213] No doubt the first act of the calculus consists in a 'depotentialisation' of the equation (for example, instead of $2ax - x^2 = y^2$ we have $dy/dx = (a - x)/y$). However, the analogue may be found in the two preceding figures where the

---

[212]Quite true; and, as far as mathematics is concerned, such a rigorous exposition has existed for more than 150 years. One wonders why a philosopher would choose to ignore it.

[213]This sentence repeats the confusion, going back to Hegel, mentioned in note 204 above.

disappearance of the *quantum* and the *quantitas* was the condition for the appearance of the element of quantitability, and disqualification the condition for the appearance of the element of qualitability. This time, following Lagrange's presentation, the depotentialisation conditions pure potentiality by allowing an evolution of the function of a variable in a series constituted by the powers of i (undetermined quantity) and the coefficients of these powers (new functions of x), in such a way that the evolution function of that variable be comparable to that of the others. The pure element of potentiality appears in the first coefficient or the first derivative, the other derivatives and consequently all the terms of the series resulting from the repetition of the same operations. The whole problem, however, lies precisely in determining this first coefficient which is itself independent of i.[214] (Deleuze 1994, pp. 174–175, italics in the original)

There is thus another part of the object which is determined by actualisation. Mathematicians ask: What is this other part represented by the so-called primitive function? In this sense, integration is by no means the inverse of differen*tia*tion[215] but, rather, forms an original process of differen*ci*ation. Whereas differen*tia*tion determines the virtual content of the Idea as problem, differen*ci*ation expresses the actualisation of this virtual and the constitution of solutions (by local integrations). Differen*ci*ation is like the second part of difference, and in order to designate the integrity or the integrality of the object we require the complex notion of different/ciation. (Deleuze 1994, p. 209, italics in the original)

---

[214]This is an extremely pedantic way to introduce Taylor series, and we doubt that this passage could be understood by anyone who did not already know the subject. Furthermore, Deleuze (as well as Hegel) bases himself on an archaic definition of "function" (namely, by its Taylor series) that goes back to Lagrange (circa 1770) but which has been superseded ever since the work of Cauchy (1821). See, for example, Boyer (1959 [1949], pp. 251–253, 267–277).

[215]See note 210 above.

These texts contain a handful of intelligible sentences—sometimes banal, sometimes erroneous—and we have commented on some of them in the footnotes. For the rest, we leave it to the reader to judge. The bottom line is: What is the point of all these mystifications about mathematical objects that have been well understood for over 150 years?

Let us look briefly at the other book "among the greatest of the great", *The Logic of Sense*, where one finds the following striking passage:

> In the first place, singularities-events correspond to heterogeneous series which are organized into a system which is neither stable nor unstable, but rather "metastable," endowed with a potential energy wherein the differences between series are distributed. (Potential energy is the energy of the pure event, whereas forms of actualization correspond to the realization of the event.) In the second place, singularities possess a process of auto-unification, always mobile and displaced to the extent that a paradoxical element traverses the series and makes them resonate, enveloping the corresponding singular points in a single aleatory point and all the emissions, all dice throws, in a single cast. In the third place, singularities or potentials haunt the surface. Everything happens at the surface in a crystal which develops only on the edges. Undoubtedly, an organism is not developed in the same manner. An organism does not cease to contract in an interior space and to expand in an exterior space—to assimilate and to externalize. But membranes are no less important, for they carry potentials and regenerate polarities. They place internal and external spaces into contact, without regard to distance. The internal and the external, depth and height, have biological value only through this topological surface of contact. Thus, even biologically, it is necessary to understand that "the deepest is the skin." The skin has at its disposal a vital and properly superficial potential energy. And just as events do not occupy the surface but rather frequent it, superficial energy is not *localized* on the surface, but is rather bound to its forma-

tion and reformation. (Deleuze 1990, pp. 103–104, italics in
the original)

Once again, this paragraph—which prefigures the style of
Deleuze's later work written in collaboration with Guattari—is
stuffed with technical terms[216]; but, apart from the banal observation that a cell communicates with the outside world through
its membrane, it is devoid of both logic and sense.

To conclude, let us quote a brief excerpt from the book
Chaosmosis, written by Guattari alone. This passage contains
the most brilliant mélange of scientific, pseudo-scientific, and
philosophical jargon that we have ever encountered; only a ge-
nius could have written it.

> We can clearly see that there is no bi-univocal corre-
> spondence between linear signifying links or archi-writing,
> depending on the author, and this multireferential, multidi-
> mensional machinic catalysis. The symmetry of scale, the
> transversality, the pathic non-discursive character of their
> expansion: all these dimensions remove us from the logic of
> the excluded middle and reinforce us in our dismissal of the
> ontological binarism we criticised previously. A machinic
> assemblage, through its diverse components, extracts its
> consistency by crossing ontological thresholds, non-linear
> thresholds of irreversibility, ontological and phylogenetic
> thresholds, creative thresholds of heterogenesis and au-
> topoiesis. The notion of scale needs to be expanded to con-
> sider fractal symmetries in ontological terms.

---

[216]For example: *singularity, stable, unstable, metastable, potential energy, singular
point, crystal, membrane, polarity, topological surface.* A defender of Deleuze might
contend that he is using these words here only in a metaphorical or philosophical
sense. But in the next paragraph, Deleuze discusses "singularities" and "singular
points" using mathematical terms taken from the theory of differential equations
(*cols, nœuds, foyers, centres*) and continues by quoting, in a footnote, a passage of a
book on differential equations that uses words like "singularity" and "singular point"
in their technical mathematical sense. See also Deleuze (1990, pp. 50, 54, 339–340n).
Deleuze is, of course, welcome to use these words in more than one sense if he likes,
but in that case he should *distinguish* between the two (or more) senses and provide
an *argument* explaining the relation between them.

What fractal machines traverse are substantial scales. They traverse them in engendering them. But, and this should be noted, the existential ordinates that they "invent" were always already there. How can this paradox be sustained? It's because everything becomes possible (including the recessive smoothing of time, evoked by René Thom) the moment one allows the assemblage to escape from energetico-spatio-temporal coordinates. And, here again, we need to rediscover a manner of being of Being—before, after, here and everywhere else—without being, however, identical to itself; a processual, polyphonic Being singularisable by infinitely complexifiable textures, according to the infinite speeds which animate its virtual compositions.

The ontological relativity advocated here is inseparable from an enunciative relativity. Knowledge of a Universe (in an astrophysical or axiological sense) is only possible through the mediation of autopoietic machines. A zone of self-belonging needs to exist somewhere for the coming into cognitive existence of any being or any modality of being. Outside of this machine/Universe coupling, beings only have the pure status of a virtual entity. And it is the same for their enunciative coordinates. The biosphere and mecanosphere, coupled on this planet, focus a point of view of space, time and energy. They trace an angle of the constitution of our galaxy. Outside of this particularised point of view, the rest of the Universe exists (in the sense that we understand existence here-below) only through the virtual existence of other autopoietic machines at the heart of other bio-mecanospheres scattered throughout the cosmos. The relativity of points of view of space, time and energy do not, for all that, absorb the real into the dream. The category of Time dissolves into cosmological reflections on the Big Bang even as the category of irreversibility is affirmed. Residual objectivity is what resists scanning by the infinite variation of points of view constitutable upon it. Imagine an autopoietic entity whose particles are constructed from galaxies. Or, conversely, a cognitivity constituted on the scale of quarks. A different panorama, another ontological consistency. The mecanosphere draws out and actualises configurations which exist amongst an infinity of others in fields of virtuality. Existential machines are at the same level as being in its intrinsic multiplicity. They are not mediated by transcendent signifiers and subsumed by a univocal ontological foundation. They are to themselves their own material of semiotic expression. Existence, as a process of deterritorialisation, is a specific inter-machinic operation which superimposes itself on the promotion of singularised existential intensities. And, I repeat, there is no generalised syntax for these deterritorialisations. Existence is not dialectical, not representable. It is hardly livable! (Guattari 1995, pp. 50–52)

Should the reader entertain any further doubts about the ubiquity of pseudo-scientific language in Deleuze and Guattari's work, he or she is invited to consult, in addition to the references given in the footnotes, pages 20–24, 32, 36–42, 50, 117–133, 135–142, 151–162, 197, 202–207, and 214–217 of *What is Philosophy?*[217], and pages 32–33, 142–143, 211–212, 251–252, 293–295, 361–365, 369–374, 389–390, 461, 469–473, and 482–490 of *A Thousand Plateaus*. These lists are by no means exhaustive. Besides, the article of Guattari (1988) on tensor calculus applied to psychology is a real gem.[218]

---

[217]This book is, in fact, densely filled with mathematical, scientific and pseudo-scientific terminology, used most of the time in a completely arbitrary way.

[218]For examples of academic articles that elaborate Deleuze and Guattari's pseudo-science, see Rosenberg (1993), Canning (1994) and the recent academic conference devoted to "DeleuzeGuattari and Matter" (University of Warwick 1997).

10. Paul Virilio

Architect and urban planner, former director of the École Spéciale d'Architecture, Paul Virilio poses questions about speed and space starting from the experience of war. For him, the mastery of time refers to power. With an astonishing erudition, which mixes space-distances and time-distances, this researcher opens up an important field of philosophical questions that he calls "dromocracy" (from the Greek *dromos:* speed).[219]

—*Le Monde (1984b, p. 195)*

The writings of Paul Virilio revolve principally around the themes of technology, communication, and speed. They contain a plethora of references to physics, particularly the theory of relativity. Though Virilio's sentences are slightly more meaningful than those of Deleuze–Guattari, what is presented as "science" is a mixture of monumental confusions and wild fantasies. Furthermore, his analogies between physics and social questions are the most arbitrary imaginable, when he does not simply become intoxicated with his own words. We confess our sympathy with many of Virilio's political and social views; but the cause is not, alas, helped by his pseudo-physics.

Let us start with a minor example of the astonishing erudition vaunted by *Le Monde*:

> Recent MEGALOPOLITAN hyperconcentration (Mexico City, Tokyo . . .) being itself the result of the increased speed of economic exchanges, it seems necessary to reconsider the importance of the notions of ACCELERATION and DECELERATION

---

[219]As Revel (1997) has noted, *dromos* does not mean "speed", but rather "course, race, running"; the Greek word for "speed" is *tachos*. Probably the error is *Le Monde*'s, because Virilio (1997, p. 22) gives the correct definition.

(what physicists call positive and negative velocities [*vitesses positive et négative selon les physiciens*]) . . . (Virilio 1995, p. 24, capitals in the original[220])

Here Virilio mixes up velocity (*vitesse*) and acceleration, the two basic concepts of kinematics (the description of motion), which are introduced and carefully distinguished at the beginning of every introductory physics course.[221] Perhaps this confusion isn't worth stressing; but for a purported specialist in the philosophy of speed, it is nonetheless a bit surprising.

Drawing inspiration from the theory of relativity, Virilio continues:

> How can we fully take in such a situation without enlisting the aid of a new type of interval, THE INTERVAL OF THE LIGHT KIND (neutral sign)? The relativistic innovation of this third interval is actually in itself a sort of unremarked cultural revelation.
>
> If the interval of TIME (positive sign) and the interval of SPACE (negative sign) have laid out the geography and history of the world through the geometric design of agrarian areas (fragmentation into plots of land) and urban areas (the cadastral system), the organization of calendars and the measurement of time (clocks) have also presided over a vast chronopolitical regulation of human societies. The very recent emergence of an interval of the third type thus signals a sudden qualitative leap, a profound mutation in the relationship between man and his surroundings.

---

[220]Translation ours. See note 221 below for a critique of the published English translations (Virilio 1993, p. 5 and 1997, p. 12).

[221]Acceleration is the *rate of change* of velocity. This confusion is systematic in Virilio's work: see, for example, Virilio (1997, pp. 31, 32, 43, 142). One of Virilio's translators (Virilio 1993, p. 5) made things worse by rendering *vitesse* as "speed" rather than "velocity". In English physics usage, "speed" designates the *length* of the velocity vector and thus can *never* be negative. The other translator (Virilio 1997, p. 12) tried to improve matters by inserting the words "vector quantities" (which do not appear in the French original) before "positive or negative velocities"; but this interpolation, while correct, leaves untouched the fundamental confusion between velocity and acceleration.

TIME (duration) and SPACE (extension) are now incon-
ceivable without LIGHT (limit-speed), the cosmological constant of the SPEED OF LIGHT . . . (Virilio 1995, p. 25; Virilio 1997,
pp. 12–13; capitals in the original)

It is true that, in the special theory of relativity, one introduces
"space-like", "time-like", and "light-like" intervals whose "invariant lengths" are respectively positive, negative, and zero
(according to the usual convention). However, these are intervals in *space-time*, which do not coincide with what we habit-
ually call "space" and "time".[222] Above all, they have nothing to
do with the "geography and history of the world" or the "chrono-
political regulation of human societies". The "very recent emergence of an interval of the third type" is nothing but a pedantic
allusion to modern telecommunications. In this passage, Virilio
shows perfectly how to package a banal observation in sophisticated terminology.

What comes next is even more surprising:

Listen to the physicist speaking about the logic of particles: "A representation is defined by a complete set of
commuting observables." [G. Cohen Tannoudji and M. Spiro,
La matière-espace-temps, Paris, Fayard, 1986.] There is no
better description of the macroscopic logic of the REAL-TIME
technologies of this sudden "teletopical commutation" that
completes and perfects what until now had been the fundamentally "topical" nature of the City of Man. (Virilio 1995,
p. 26, capitals in the original[223])

The sentence "A representation is defined by a complete set of
commuting observables" is a rather common technical expression in *quantum mechanics* (not in relativity). It has nothing to
do with "real time" or with any "macroscopic logic" (quite the

---

[222]The book by Taylor and Wheeler (1966) gives a beautiful introduction to the notion
of space-time interval.

[223]Translation ours. See note 224 below for a critique of the published English
translations (Virilio 1993, p. 6 and 1997, p. 13).

contrary, it refers to *micro*physics), much less with "teletopical commutation" or the "City of Man". But above all, in order to understand the precise meaning of this sentence, one needs to have studied physics and mathematics seriously for several years. We find it incredible that Virilio could *consciously* copy a sentence that he manifestly does not understand, add to it a completely arbitrary comment, and still be taken seriously by editors, commentators, and readers.[224,225]

Virilio's works are overflowing with this pseudo-scientific verbiage.[226] Here is another example:

> What happens to the transparence of air, of water, of glass— one could say, of the "real space" of things surrounding us— when the *interface* of "real time" takes over from the classic "interval," and when *distance* suddenly gives way to the *power* of transmission and instantaneous reception? . . . Transparence is no longer composed of light rays (solar or electric) but instead of elemental particles (electrons and photons) that are transmitted at the speed of light. (Virilio 1989, p. 129; Virilio 1990, p. 107; italics in the original)

---

[224]Virilio's English translators—who can hardly be expected to possess a technical knowledge of physics—have likewise made a hash of this sentence. One rendered it as "A representation is defined by a sum of observables that are flickering back and forth" (Virilio 1993, p. 6), while the other came up with "A display is defined by a complete set of observables that commutate" (Virilio 1997, p. 13).

[225]Here is how a book containing this essay of Virilio's was lauded in an American journal of literary studies:

> *Re-thinking Technologies* constitutes a significant contribution to the analysis of techno-cultures today. It will definitely contradict those who still think that postmodernity is merely a fashionable term or an empty fad. The nagging opinion that cultural and critical theory is "too abstract," hopelessly removed from reality, devoid of ethical values and above all incompatible with *erudition, systematic thinking, intellectual rigor* and creative criticism, will simply be pulverized. . . . This collection assembles some of the most recent and fresh work by leading culture critics and theoreticians of the arts and sciences, such as Paul Virilio, Félix Guattari, . . . (Gabon 1994, pp. 119–120, emphasis added)

It is amusing to see the reviewer's misunderstandings as he tries to understand (and thinks he understands) Virilio's fantasies concerning relativity. We fear that more cogent arguments will be required to pulverize our own "nagging opinions".

[226]Particularly *L'Espace critique* (1984), *L'Inertie polaire* (1990) and *La Vitesse de libération* (1995). The first of these is translated as *The Lost Dimension* (1991), and the third as *Open Sky* (1997).

For what it's worth, electrons, unlike photons, have a non-zero mass and thus *cannot* move at the speed of light, precisely because of the theory of relativity of which Virilio seems so fond.

In what follows, Virilio continues to throw around scientific terminology, supplemented by his own inventions (*teletopology, chronoscopy*):

> This displacement of the direct transparence of a material is due primarily . . . to the effective use of *undulatory optics* alongside classic *geometric optics*. In the same way that alongside Euclidean geometry we find a non-Euclidean, topological geometry, the passive optics of the geometry of camera lenses and telescopes is accompanied by the active optics of the *teletopology* of optoelectric waves.
>
> . . . Traditional chronology—future, present, past—has been succeeded by CHRONOSCOPY—underexposed, exposed, overexposed. The interval of the TIME genre (the positive sign) and the interval of the SPACE genre (the negative sign, with the same name as the inscription surface of film) are inscribed only by LIGHT, that interval of the third genre in which the *zero* sign means absolute speed.
>
> The exposure time of the photographic plate is therefore merely time's (space-time's) exposure of its photosensitive material to the light of speed, which is to say, finally, to the frequency of photon-carrying waves. (Virilio 1989, p. 129; Virilio 1990, pp. 108–109, 115; italics and capitals in the original)

This mélange of optics, geometry, relativity, and photography needs no comment.

Let us round off our reading of Virilio's writings on speed with this little marvel:

> Remember that the dromospheric space, space-speed, is physically described by what is called the "logistic equation," the

result of the product of the mass displaced by the speed of its displacement, MxV. (Virilio 1984, p. 176; Virilio 1991, p. 136[227])

The logistic equation is a differential equation studied in population biology (among other fields); it is written $dx/dt = \lambda x(1-x)$ and was introduced by the mathematician Verhulst (1838). It has nothing to do with $M \times V$. In Newtonian mechanics, $M \times V$ is called "momentum"; in relativistic mechanics, $M \times V$ does not arise at all. The *dromospheric space* is a Virilian invention.

Of course, no work in this genre would be complete without an allusion to Gödel's theorem:

> This drifting of figures and geometric figuring, this irruption of dimensions and transcendental mathematics, leads us to the promised surrealist peaks of scientific theory, peaks that culminate in Gödel's theorem: the existential proof, a method that mathematically proves the existence of an object without producing that object . . . (Virilio 1991, p. 66)

In reality, existential proofs are much older than Gödel's work; and the proof of his theorem is, by contrast, completely constructive: it exhibits a proposition that is neither provable nor refutable in the system under consideration (provided the system is consistent).[228]

And, to top it all off:

> When depth of time replaces depths of sensible space; when the commutation of interface supplants the delimitation of surfaces; when transparence re-establishes appearances; then we begin to wonder whether that which we insist on calling *space* isn't actually *light*, a subliminary, para-optical light of which sunlight is only one phase or reflection. This light occurs in a duration measured in instantaneous time ex-

---

[227]We have corrected a typographical error in the translation, in which *"espace dromosphérique"* was rendered as "dromospheric sphere" rather than "dromospheric space".

[228]See, for example, Nagel and Newman (1958).

posure rather than the historical and chronological passage of time. The time of this instant without duration is "exposure time", be it over- or underexposure. Its photographic and cinematographic technologies already predicted the existence and the time of a continuum stripped of all physical dimensions, in which the quantum of energetic action and the punctum of cinematic observation have suddenly become the last vestiges of a vanished morphological reality. Transferred into the eternal present of a relativity whose topological and teleological thickness and depth belong to this final measuring instrument, this speed of light possesses one direction, which is both its size and dimension and which propagates itself at the same speed in all radial directions that measure the universe. (Virilio 1984, p. 77; Virilio 1991, pp. 63–64; italics in the original)

This paragraph—which in the French original is a single 193-word sentence, whose "poetry" is unfortunately not fully captured by the translation—is the most perfect example of diarrhea of the pen that we have ever encountered. And as far as we can see, it means precisely nothing.

11. Gödel's Theorem and Set Theory: Some Examples of Abuse

Ever since Gödel showed that there does not exist a proof of the consistency of Peano's arithmetic that is formalizable within this theory (1931), political scientists had the means for understanding why it was necessary to mummify Lenin and display him to the "accidental" comrades in a mausoleum, at the Center of the National Community.

—*Régis Debray, Le Scribe (1980, p. 70)*

In applying Gödel's theorem to questions of the closed and the open, as they relate to sociology then, with one gesture Régis Debray recapitulates and concludes the history and the work of the previous 200 years.

—*Michel Serres, A History of Scientific Thought (1995, p. 452)*

Gödel's theorem is an inexhaustible source of intellectual abuses: we have already encountered examples in Kristeva and Virilio, and a whole book could easily be written on the subject. In this chapter we shall give some rather extraordinary examples in which Gödel's theorem and other concepts taken from the foundations of mathematics are extrapolated in a totally arbitrary way to the social and political domain.

The social critic Régis Debray devotes one chapter of his theoretical work, *Critique of Political Reason*, to explain that "collective madness finds its ultimate foundation in a logical axiom that is itself without foundation: *incompleteness*".[229] This

---

[229]Debray (1981, p. 10).

"axiom" (also called "thesis" or "theorem") is introduced in a rather bombastic fashion:

> The 'secret' of our collective miseries, of the a priori condition of any political history, past, present or future, may be stated in a few simple, even childish words. If we bear in mind that surplus labour and the unconscious can both be defined in a single sentence (and that, in the physical sciences, the equation for general relativity can be stated in three letters), there is no danger of confusing simplicity with over-simplification. The secret takes the form of a logical law, an extension of Gödel's theorem: *there can be no organized system without closure and no system can be closed by elements internal to that system alone.* (Debray 1983, pp. 169–170, italics in the original)

Let us pass over the allusion to general relativity. What is more serious is the invocation of Gödel's theorem, which concerns the properties of certain formal systems in mathematical logic, to explain the "secret of our collective miseries". There is quite simply no logical relationship between this theorem and questions of sociology.[230]

Nevertheless, the conclusions that Debray draws from his "extension of Gödel's theorem" are rather spectacular. For example:

> Just as it would be a biological contradiction for an individual to give birth to himself (integral cloning as biological aporia?), the government of a collective by itself—*verbi gratia* 'of the people by the people'—is a logically contradictory opera-

---

[230] The text quoted here is relatively old; but one finds the same idea in *Media Manifestos* (1994, p. 12 and 1996a, p. 4). Subsequently, however, Debray seems to have retreated to a more prudent position: in a recent lecture (Debray 1996b) he admits that "Gödelitis is a widespread disease" (p. 6) and that "extrapolating a scientific result, and generalizing it outside of its specific field of relevance, can lead . . . to gross errors" (p. 7); he says also that his use of Gödel's theorem is intended as "simply metaphorical or isomorphic" (p. 7).

tion ('generalized workers' control' as political aporia). (Debray 1983, p. 177)

And likewise:

> It is quite natural that there should be something irrational about groups, for if there were not, there would be no groups. It is positive that there should be something mystical about them, because a demystified society would be a pulverized society. (Debray 1983, p. 176)

According to Debray, therefore, neither a government of the people by the people nor a demystified society are possible, and this apparently for strictly *logical* reasons.

But if the argument were valid, one might as well use it to prove the existence of God, as suggested by the following statement:

> Incompleteness stipulates that a set cannot, by definition, be a substance in the Spinozist sense: something which exists in itself and is conceived by itself. It requires a cause (to engender it) and it is not its own cause. (Debray 1983, p. 177)

Nevertheless, Debray denies the existence of God (p. 176), without explaining why it would not be an equally "logical" consequence of his "theorem".

The bottom line is that Debray never explains what role Gödel's theorem is supposed to play in his argument. If he wants to employ it directly in reasoning about social organization, then he is simply wrong. If, on the other hand, Gödel's theorem is intended to serve merely as an analogy, then it could be suggestive but certainly not demonstrative. To support his sociological and historical theses, he would have to supply arguments dealing with humans and their social behavior, not mathematical logic.

Gödel's theorem will still be true in ten thousand years or a million years, but no one can say what human society will look like so far in the future. The invocation of this theorem thus gives the appearance of an "eternal" quality to theses that are, at best, valid in a given context and at a given time. Indeed, the allusion to the "biological contradiction" supposedly inherent in "integral cloning" looks nowadays a bit out-of-date—which shows that one must be careful when "applying" Gödel's theorem.

Since this idea of Debray does not seem terribly impressive, we were quite surprised to see it elevated to a "Gödel-Debray principle" by the renowned philosopher Michel Serres[231], who explains that

> Régis Debray applies or discovers as applicable to social groups the incompleteness theorem valid for formal systems, and shows that societies can only organize themselves on the express condition that they are founded on something other than themselves, outside their own definition or border. They cannot be sufficient in themselves. He calls this foundation religious. With Gödel he completes the work of Bergson, whose *Les Deux Sources de la morale et de la religion* [*The Two Sources of Morality and Religion*] differentiated between closed and open societies. No, he says, internal coherence is guaranteed by the external: the group only closes if it is open. Saints, geniuses, heroes, paragons and all sorts of champions do not break institutions but make them possible. (Serres 1995, pp. 449–450)

He continues:

> Since Bergson, the most notable historians have copied from the *Two Sources* . . . Far from copying a model, as they do, Régis Debray solves a problem. Where historians describe the crossing or transgression of social or conceptual limits, without understanding them, because they have borrowed their model ready-made from Bergson, which Bergson con-

---

[231]Serres (1995, p. 451). See also Dhombres (1994, p. 195) for a critical remark on this "principle".

structed on the basis of Carnot and thermodynamics, Régis Debray has constructed his own, and has therefore grasped a new model, based on Gödel and on logical systems.

This decisive contribution from Gödel and Debray frees us from ancient models and from their repetition. (Serres 1995, p. 450)

Serres goes on to apply the "Gödel–Debray principle" to the history of science[232], where it is as irrelevant as it is in politics.

Our last example is reminiscent of Sokal's parody, where he plays on the word "choice" to forge an absurd link between the axiom of choice in mathematical set theory[233] and the movement for abortion rights. He goes so far as to invoke Cohen's theorem, which shows that the axiom of choice and the continuum hypothesis[234] are independent (in the technical meaning of this word in logic) of the other axioms of set theory, to claim that conventional set theory is insufficient for a "liberatory" mathematics. Here again, one finds a completely arbitrary leap from the foundations of mathematics to political considerations.

Since this passage is one of the most openly ridiculous in the parody, we were rather surprised to find similar ideas put forward in utter seriousness—or so it appears—by the philosopher Alain Badiou (in texts that we emphasize are rather old). In *Theory of the Subject* (1982), Badiou happily throws together politics, Lacanian psychoanalysis, and mathematical set theory. The following excerpt from the chapter entitled "The logic of excess" gives an idea of the book's flavor. After a brief discussion

---

[232]Where one finds this gem: speaking of the Ancien Régime, Serres writes that "the clergy occupied a very precise place in society. Dominant and dominated, neither dominated nor dominant, this place, within each dominant or dominated class, belonged to neither one nor the other, to neither the dominated nor the dominant." (Serres 1995, p. 453)

[233]See p. 44 above for a brief explanation of the axiom of choice.

[234]See note 41 above for a brief explanation of the continuum hypothesis.

on the situation of immigrant workers, Badiou refers to the continuum hypothesis and continues (pp. 282–283):

> What is at stake here is nothing less than the fusion of algebra (ordered succession of cardinals) and topology (excess of the partitive over the elementary). The truth of the continuum hypothesis would make law [*ferait loi*] of the fact that the excess in the multiple has no other assignment than the occupation of the empty place, than the existence of the nonexistent proper of the initial multiple. There would be this maintained filiation of coherence, that what exceeds internally the whole does not go beyond naming the limit point of this whole.
>
> But the continuum hypothesis is not provable.
>
> Mathematical triumph of politics over trade-union realism.[235]

One cannot help but wonder whether a few paragraphs were inadvertently omitted between the last two sentences of this quote; but no such luck, the jump between mathematics and politics is as abrupt as it appears.[236]

---

[235]The French Maoist discourse of the late 1960s insisted on a sharp opposition between "politics", which was supposed to be put in the commanding position, and trade unionism.

[236]For what it's worth, the "mathematics" in this paragraph are also rather meaningless.

12. Epilogue

In this last chapter, we shall address some general questions—historical, sociological, and political—that arise naturally from a reading of the texts quoted in this book. We shall limit ourselves to explaining our point of view, without justifying it in detail. It goes without saying that we claim no special competence in history, sociology, or politics; and what we have to say must, in any case, be understood as conjectures rather than as the final word. If we do not simply remain silent on these questions, it is principally to avoid having ideas attributed to us against our will (as has already been done) and to show that our position on many issues is quite moderate.

Over the past two decades, much ink has been spilled about postmodernism, an intellectual current that is supposed to have replaced modern rationalist thought.[237] However, the term "postmodernism" covers an ill-defined galaxy of ideas—ranging from art and architecture to the social sciences and philosophy—and we have no wish to discuss most of these areas.[238] Our focus is limited to certain intellectual aspects of

---

[237]We do not want to get involved in terminological disputes about the distinctions between "postmodernism", "poststructuralism", and so forth. Some writers use the term "poststructuralism" (or "anti-foundationalism") to denote a particular collection of philosophical and social theories, and "postmodernism" (or "postmodernity") to denote a wider set of trends in contemporary society. For simplicity, we shall use the term "postmodernism", while emphasizing that we shall be concentrating on the philosophical and intellectual aspects and that the validity or invalidity of our arguments can in no way depend on the use of a word.

[238]Indeed, we have no strong views on postmodernism in art, architecture, or literature.

postmodernism that have had an impact on the humanities and the social sciences: a fascination with obscure discourses; an epistemic relativism linked to a generalized skepticism toward modern science; an excessive interest in subjective beliefs independently of their truth or falsity; and an emphasis on discourse and language as opposed to the facts to which those discourses refer (or, worse, the rejection of the very idea that facts exist or that one may refer to them).

Let us start by recognizing that many "postmodern" ideas, expressed in a moderate form, provide a needed correction to naive modernism (belief in indefinite and continuous progress, scientism, cultural Eurocentrism, etc.). What we are criticizing is the radical version of postmodernism, as well as a number of mental confusions that are found in the more moderate versions of postmodernism and that are in some sense inherited from the radical one.[239]

We shall begin by considering the tensions that have always existed between the "two cultures" but that seem to have worsened during the last few years, as well as the conditions for a fruitful dialogue between the humanities and social sciences and the natural sciences. We shall then analyze some of the intellectual and political sources of postmodernism. Finally, we shall discuss the negative aspects of postmodernism for both culture and politics.

For a Real Dialogue Between the "Two Cultures"

Interdisciplinarity seems to be the order of the day. Though some people worry that the dilution of specialization may lead to a decline in the standards of intellectual rigor, the insights that one field of thought can bring to another cannot be ignored. By no means do we wish to inhibit interaction between the

---

[239]See also Epstein (1997) for a useful distinction between the "weak" and "strong" versions of postmodernism.

mathematico-physical sciences and the human sciences; rather, our aim is to emphasize some preconditions we see as necessary for a real dialogue.

Over the past few years, it has become fashionable to talk about a so-called "science war".[240] But this phrase is quite unfortunate. Who is waging war, and against whom?

Science and technology have long been the subject of philosophical and political debates: on nuclear weapons and nuclear energy, the human genome project, sociobiology, and many other subjects. But these debates in no way constitute a "science war". Indeed, many different reasonable positions in these debates are advocated by scientists and non-scientists alike, using scientific and ethical arguments that can be rationally evaluated by all the people involved, whatever their profession.

Unfortunately, some recent developments may lead one to fear that something completely different is going on. For example, researchers in the social sciences can legitimately feel threatened by the idea that neurophysiology and sociobiology will replace their disciplines. Similarly, people working in the natural sciences may feel under attack when Feyerabend calls science a "particular superstition"[241] or when some currents in the sociology of science give the impression of placing astronomy and astrology on the same footing.[242]

In order to alleviate these fears, it is worth distinguishing between the claims made for research programmes, which tend to be grandiose, and the actual accomplishments, which are gen-

---

[240]This expression was apparently first used by Andrew Ross, one of the editors of *Social Text*, who asserted (rather tendentiously) that

> the Science Wars [are] a second front opened up by conservatives cheered by the successes of their legions in the holy Culture Wars. Seeking explanations for their loss of standing in the public eye and the decline in funding from the public purse, conservatives in science have joined the backlash against the (new) usual suspects—pinkos, feminists and multiculturalists. (Ross 1995, p. 346)

Later, the phrase was used as the title of the special issue of *Social Text* in which Sokal's hoax article appeared (Ross 1996).

[241]See Feyerabend (1975, p. 308).

[242]See, for example, Barnes, Bloor and Henry (1996, p. 141); and for a cogent critique, see Mermin (1998).

erally rather modest. The basic principles of chemistry are today entirely based on quantum mechanics, hence on physics; and yet, chemistry as an autonomous discipline has not disappeared (even if some parts of it have gotten closer to physics). Likewise, if one day the biological bases of our behavior were sufficiently well understood to serve as a foundation for the study of human beings, there would be no reason to fear that the disciplines we now call "social sciences" would somehow disappear or become mere branches of biology.[243] In a similar way, scientists have nothing to fear from a realistic historical and sociological view of the scientific enterprise, provided that a certain number of epistemological confusions are avoided.[244]

Let us therefore put aside the "science war", and see what kind of lessons can be drawn from the texts cited in this book concerning the relationship between the natural and the human sciences.[245]

1. *It's a good idea to know what one is talking about.* Anyone who insists on speaking about the natural sciences—and nobody is forced to do so—needs to be well-informed and to avoid making arbitrary statements about the sciences or their epistemology. This may seem obvious, but as the texts gathered in this book demonstrate, it is all too often ignored, even (or especially) by renowned intellectuals.

Obviously, it is legitimate to think philosophically about the content of the natural sciences. Many concepts used by scientists—such as the notions of law, explanation, and causality—contain hidden ambiguities, and philosophical reflection can help to clarify the ideas. But, in order to address these subjects meaningfully, one has to understand the relevant scientific the-

---

[243]Which is not to say, of course, that they would not be profoundly modified, as chemistry was.

[244]See Sokal (1998) for an extensive, though by no means exhaustive, list of what we see as valid tasks for the history and sociology of science.

[245]We emphasize that what follows is *not* intended as a comprehensive list of the conditions for a fruitful dialogue between the natural and the human sciences, but simply as a reflection on the lessons to be drawn *from the texts cited in this book.* Many other criticisms can, of course, be made of *both* the natural and the human sciences, but they are beyond the scope of the present discussion.

ories at a rather deep and inevitably technical level[246]; a vague understanding, at the level of popularizations, won't suffice.

2. *Not all that is obscure is necessarily profound.* There is a huge difference between discourses that are difficult because of the inherent nature of their subject and those whose vacuity or banality is carefully hidden behind deliberately obscure prose. (This problem is by no means specific to the humanities or social sciences; many articles in physics and mathematics use a language more complicated than is strictly necessary.) Of course, it is not always easy to determine which kind of difficulty one is facing; and those who are accused of using obscure jargon frequently reply that the natural sciences also use a technical language that can be mastered only after many years of study. Nevertheless, it seems to us that there are some criteria that can be used to help distinguish between the two sorts of difficulty. First, when the difficulty is genuine, it is usually possible to explain in simple terms, at some rudimentary level, what phenomena the theory is examining, what are its main results, and what are the strongest arguments in its favor.[247] For example, although neither of us has any training in biology, we are able to follow, at some basic level, developments in that field by reading good popular or semi-popular books. Second, in these cases there is a clear path—possibly a long one—that will lead to a deeper knowledge of the subject. By contrast, some obscure discourses give the impression that the reader is being asked to make a qualitative jump, or to undergo an experience similar to a revelation, in order to understand them.[248] Again, one cannot help being reminded of the emperor's new clothes.[249]

---

[246]As positive examples of this attitude, let us mention, among others, the works of Albert (1992) and Maudlin (1994) on the foundations of quantum mechanics.

[247]To give just a few examples, let us mention Feynman (1965) in physics, Dawkins (1986) in biology, and Pinker (1995) in linguistics. We do not necessarily agree with everything these authors say, but we consider them models of clarity.

[248]For similar observations, see the remarks of Noam Chomsky quoted by Barsky (1997, pp. 197–198).

[249]We don't want to be unduly pessimistic about the probable response to our book, but let us note that the story of the emperor's new clothes ends as follows: "And the chamberlains went on carrying the train that wasn't there."

3. *Science is not a "text"*. The natural sciences are not a mere reservoir of metaphors ready to be used in the human sciences. Non-scientists may be tempted to isolate from a scientific theory some general "themes" that can be summarized in few words such as "uncertainty", "discontinuity", "chaos", or "nonlinearity" and then analyzed in a purely verbal manner. But scientific theories are not like novels; in a scientific context these words have specific meanings, which differ in subtle but crucial ways from their everyday meanings, and which can only be understood within a complex web of theory and experiment. If one uses them only as metaphors, one is easily led to nonsensical conclusions.[250]

4. *Don't ape the natural sciences*. The social sciences have their own problems and their own methods; they are not obliged to follow each "paradigm shift" (be it real or imaginary) in physics or biology. For example, although the laws of physics at the atomic level are expressed today in a probabilistic language, deterministic theories can nevertheless be valid (to a very good approximation) at other levels, for example in fluid mechanics or even possibly (and yet more approximately) for certain social or economic phenomena. Conversely, even if the fundamental physical laws were perfectly deterministic, our ignorance would force us to introduce a great number of probabilistic models in order to study phenomena at other levels, like gases or societies. Besides, even if one adopts a reductionist *philosophical* attitude, one is by no means obliged to pursue reductionism as a *methodological* prescription.[251] In practice, there are so many orders of magnitude separating atoms from fluids, brains, or societies that vastly different models and methods are quite naturally employed in each realm, and establishing a link between

---

[250]For example, a sociologist friend asked us, not unreasonably: Isn't it contradictory for quantum mechanics to exhibit both "discontinuity" and "interconnectedness"? Aren't these properties opposites? The brief answer is that these properties characterize quantum mechanics *in very specific senses*—which require a mathematical knowledge of the theory to be properly understood—and that, *in these senses*, the two notions do not contradict one another.

[251]See, for example, Weinberg (1992, chapter III) and Weinberg (1995).

these levels of analysis is not necessarily the most urgent task. In other words, the type of approach in each domain of research should depend upon the specific phenomena under investigation. Psychologists, for example, do not need to invoke quantum mechanics to maintain that *in their field* "the observer affects the observed"; this is a truism, irrespective of the behavior of electrons or atoms.

Moreover, there are so many phenomena, even in physics, that are imperfectly understood, at least for the time being, that there is no reason to try to imitate the natural sciences when dealing with complex human problems. It is perfectly legitimate to turn to intuition or literature in order to obtain some kind of nonscientific understanding of those aspects of human experience that cannot, at least at present, be tackled more rigorously.

5. *Be wary of argument from authority.* If the human sciences want to benefit from the undeniable successes of the natural sciences, they need not do so by directly extrapolating technical scientific concepts. Instead, they could get some inspiration from the best of the natural sciences' *methodological* principles, starting with this one: to evaluate the validity of a proposition on the basis of the facts and reasoning supporting it, without regard to the personal qualities or social status of its advocates or detractors.

This is, of course, only a *principle;* it is far from universally honored in practice, even in the natural sciences. Scientists are, after all, human beings and are not immune to fashion or to the adulation of geniuses. Nevertheless, we have inherited from the "epistemology of the Enlightenment" a totally justified mistrust toward the exegesis of sacred texts (and texts that are not religious in the traditional sense may very well fulfill that role) as well as toward argument from authority.

We met in Paris a student who, after having brilliantly finished his undergraduate studies in physics, began reading philosophy and in particular Deleuze. He was trying to tackle *Difference and Repetition.* Having read the mathematical excerpts examined here (pp. 161–164), he admitted he couldn't see what Deleuze was driving at. Nevertheless, Deleuze's reputation for profundity was so strong that he hesitated to draw the natural conclusion: that if someone like himself, who had studied calculus for several years, was unable to understand these texts, allegedly about calculus, it was probably because they didn't make much sense. It seems to us that this example should have encouraged the student to analyze more critically the rest of Deleuze's writings.

6. *Specific skepticism should not be confused with radical skepticism.* It is important to distinguish carefully between two different types of critiques of the sciences: those that are opposed to a particular theory and are based on specific arguments, and those that repeat in one form or another the traditional arguments of radical skepticism. The former critiques can be interesting but can also be refuted, while the latter are irrefutable but uninteresting (because of their universality). And it is crucial not to mix the two sorts of arguments: for if one wants to contribute to science, be it natural or social, one must abandon radical doubts concerning the viability of logic or the possibility of knowing the world through observation and/or experiment. Of course, one can always have doubts about a specific theory. But general skeptical arguments put forward to support those doubts are irrelevant, precisely because of their generality.

7. *Ambiguity as subterfuge.* We have seen in this book numerous ambiguous texts that can be interpreted in two different ways: as an assertion that is true but relatively banal, or as one that is radical but manifestly false. And we cannot help thinking that, in many cases, these ambiguities are deliberate. Indeed, they offer a great advantage in intellectual battles: the radical interpretation can serve to attract relatively inexperienced listeners or readers; and if the absurdity of this version is exposed, the author can always defend himself by claiming to have been misunderstood, and retreat to the innocuous interpretation.

How Did We Get Here?

In the debates that followed the publication of the *Social Text* parody, we were often asked: How and why did the intellectual trends that you are criticizing develop? This is a very complicated question belonging to the history and sociology of ideas, to which we certainly do not claim to have a definitive answer. We would like, rather, to put forward some possible answers, while emphasizing both their conjectural nature and their incompleteness (there are undoubtedly other elements that we have underestimated or missed entirely). Moreover, as always in this kind of complex social phenomenon, there is a mixture of very diverse causes. In this section we shall limit ourselves to the intellectual sources of postmodernism and relativism, leaving the political aspects for the next section.

1. *Neglect of the empirical.* For a long time, it has been fashionable to denounce "empiricism"; and if that word denotes an allegedly fixed method for extracting theories from facts, we can only agree. Scientific activity has always involved a complex interplay between observation and theory, and scientists have known that for a long time.[252] So-called "empiricist" science is a caricature belonging to bad schoolbooks.

Nevertheless, our theories about the physical or social world need to be justified in one way or another; and if one eschews apriorism, argument from authority, and reference to "sacred" texts, there is not much left besides the systematic test of theory by observations and experiments. One need not be a strict Popperian to realize that any theory must be supported, at least indirectly, by empirical evidence in order to be taken seriously.

Some of the texts cited in this book completely disregard the empirical aspect of science and concentrate exclusively on language and theoretical formalism. They give the impression

---

[252]For a good illustration of the complexity of the interaction between observation and theory, see Weinberg (1992, chapter V) and Einstein (1949).

that a discourse becomes "scientific" as soon as it seems superficially coherent, even if it is never subjected to empirical tests. Or, worse, that it is sufficient to throw mathematical formulae at problems in order to make progress.

2. *Scientism in the social sciences.* This point may seem bizarre: Isn't scientism the sin of physicists and biologists who seek to reduce everything to matter in motion, natural selection, and DNA? Yes and no. Let us define "scientism", for the purposes of this discussion, as the illusion that simplistic but supposedly "objective" or "scientific" methods will allow us to solve very complex problems (other definitions are certainly possible). The difficulty that constantly arises when one succumbs to such illusions is that important parts of reality are forgotten simply because they fail to fit within the framework that was posed *a priori*. Sadly, examples of scientism are abundant in the social sciences: one can cite, among others, certain currents within quantitative sociology, neoclassical economics, behaviorism, psychoanalysis, and Marxism.[253] Often what happens is that people start with a set of ideas having some validity in a given domain and, instead of trying to test them and refine them, they extrapolate them unreasonably.

Unfortunately, scientism has often been confused—by its supporters as well as by its detractors—with the scientific attitude itself. As a result, the entirely justified reaction against scientism in the social sciences has sometimes led to an equally unjustified reaction against science as such—and this on the part of both the ex-partisans and ex-opponents of the old scientisms. For example, in France after May 1968, the reaction against the scientism of certain rather dogmatic strains of structuralism and Marxism was one factor (among many others) that led to the emergence of postmodernism (the "incredulity toward metanarratives", to quote Lyotard's famous catchword[254]).

---

[253]More recent, and even more extreme, examples of scientism can be found in the alleged "applications" of the theories of chaos, complexity, and self-organization to sociology, history, and business management.

[254]Lyotard (1984, p. xxiv).

A similar evolution occurred, in the 1990s, among some intellectuals in the former Communist countries: for instance, the Czech president Václav Havel wrote that

> The fall of Communism can be regarded as a sign that modern thought—based on the premise that the world is objectively knowable, and that the knowledge so obtained can be absolutely generalized—has come to a final crisis. (Havel 1992)

(One wonders why a renowned thinker such as Havel is incapable of making the elementary distinction between the scientific worldview and the Communist regimes' unjustified *claim* to possess a "scientific" theory of human history.)

When one combines neglect of the empirical side with a good deal of scientistic dogmatism, one can be led into the worst lucubrations, of which we have seen all too many examples. But one can alternatively fall into a sort of discouragement: since such and such (simplistic) method, to which one had dogmatically adhered, does not work, therefore nothing works, all knowledge is impossible or subjective, etc. And so one passes easily from the climate of the 1960s and 1970s to postmodernism. But it is based on a misidentification of the source of the problem.

One recent avatar of scientism in the social sciences is, paradoxically, the "strong programme" in the sociology of science. To try to explain the content of scientific theories without taking into account, even in part, the rationality of scientific activity is to eliminate *a priori* an element of reality and, it seems to us, to deprive oneself of any possibility of effectively understanding the problem. To be sure, every scientific study must make simplifications and approximations; and the approach of the "strong programme" would be legitimate if its advocates were to provide empirical or logical arguments showing that the neglected aspects are indeed of marginal importance for understanding the phenomena in question. But no such arguments are given; the principle is posed *a priori*. In reality, the strong programme is trying to make a virtue of (apparent) necessity: since it is difficult for sociologists to study the internal rationality of the natural sciences, it is declared "scientific" to ignore it. It is like trying to complete a puzzle when one knows that half the pieces are missing.

We believe that the scientific attitude, understood very broadly—as a respect for the clarity and logical coherence of theories, and for the confrontation of theories with empirical evidence—is as relevant in the social sciences as it is in the natural sciences. But one must be very prudent toward claims of scientificity in the social sciences; this holds also (or even especially) for the currently dominant trends in economics, sociology, and psychology. The problems addressed by the social sciences are extremely complex, and the empirical evidence supporting their theories is often quite weak.

3. *The prestige of the natural sciences.* There is no doubt that the natural sciences enjoy an enormous prestige, even among their detractors, because of their theoretical and practical successes. Scientists sometimes abuse this prestige by displaying an unjustified feeling of superiority. Moreover, well-known scientists, in their popular writings, often put forward speculative ideas as if they were well-established, or extrapolate their results far beyond the domain where they have been verified. Finally, there is a damaging tendency—exacerbated, no doubt, by the demands of marketing—to see a "radical conceptual revolution" in each innovation. All these factors combined give the educated public a distorted view of scientific activity.

But it would be demeaning to philosophers, psychologists, and sociologists to suggest that they are defenseless in the face of such scientists, and that the abuses exposed in this book are somehow unavoidable. It is obvious that no one, and in particular no scientist, forced Deleuze or Lacan to write as they do. One can perfectly well be a psychologist or a philosopher and either speak about the natural sciences knowing what one is talking about, or else not speak about them and concentrate on other things.

4. *The social sciences' "natural" relativism.* In certain branches of the social sciences, notably in anthropology, a certain "relativistic" attitude is methodologically natural, especially when one is studying tastes or customs: the anthropologist is seeking to understand these customs' role in a given society, and it is difficult to see what she would gain by dragging into her research her own aesthetic preferences. Similarly, when studying certain cognitive aspects of a culture, such as the social role of the culture's cosmological beliefs, the anthropologist is not principally concerned with knowing whether those beliefs are true or false.[255]

However, this reasonable methodological relativism has sometimes led, through confusions of thought and language, to a radical cognitive relativism: namely, the claim that assertions of fact—be they traditional myths or modern scientific theories—can be considered true or false only "relative to a particular culture". But this amounts to confusing the psychological and social functions of a system of thought with its cognitive value, and to ignoring the strength of the empirical arguments that can be put forward in favor of one system of thought over another.

Here is a concrete example of such a confusion: There are at least two competing theories concerning the origin of Native American populations. The scientific consensus, based on extensive archaeological evidence, is that humans first came to the Americas from Asia around 10–20,000 years ago, crossing the Bering Strait. On the other hand, many Native American creation accounts hold that native peoples have always lived in the

---

[255]This last question is nevertheless rather subtle. All beliefs, even mythical ones, are constrained, at least in part, by the phenomena to which they refer. And, as we showed in Chapter 4, the "strong programme" in the sociology of science, which is a kind of anthropological relativism applied to contemporary science, goes astray precisely because it neglects this latter aspect, which plays a crucial role in the natural sciences.

Americas, ever since their ancestors emerged onto the surface of the earth from a subterranean world of spirits. And a report in the *New York Times* (22 October 1996) observed that many archaeologists, "pulled between their scientific temperaments and their appreciation for native culture . . . have been driven close to a postmodern relativism in which science is just one more belief system." For example, Roger Anyon, a British archaeologist who has worked for the Zuni people, was quoted as saying that "science is just one of many ways of knowing the world. . . . [The Zunis' world view is] just as valid as the archeological viewpoint of what prehistory is about."[256]

Perhaps Dr. Anyon was misquoted[257], but one does hear this type of assertion rather frequently nowadays, and we would like to analyze it. Note first that the word "valid" is ambiguous: is it intended in a cognitive sense, or in some other sense? If the latter, we have no objection; but the reference to "knowing the world" suggests the former. Now, both in philosophy and in everyday language, there is a distinction between *knowledge* (understood, roughly, as justified true belief) and mere *belief*; that is why the word "knowledge" has a positive connotation, while "belief" is neutral. What, then, does Anyon mean by "knowing the world"? If he intends the word "knowing" in its traditional sense, then his assertion is simply false: the two theories in question are mutually incompatible, so they cannot both be true (or even approximately true).[258] If, on the other hand, he

---

[256]Johnson (1996, p. C13). A more detailed exposition of Anyon's views can be found in Anyon *et al.* (1996).

[257]But probably not, because essentially identical views are expressed in Anyon *et al.* (1996).

[258]During a debate at New York University, where this example was mentioned, many people seemed not to understand or accept this elementary remark. The problem presumably comes, at least in part, from the fact that they have redefined "truth" as a belief that is "locally accepted as such" or else as an "interpretation" that fulfills a given psychological and social role. It is difficult to say what shocks us the most: someone who believes that the creationist myths are *true* (in the usual sense of the word) or someone who adheres systematically to this redefinition of the word "true". For a more detailed discussion of this example and in particular of the possible meanings of the word "valid", see Boghossian (1996).

is simply noting that different people have different beliefs, then his assertion is true (and banal), but it is misleading to employ the success-word "knowledge".[259]

Most likely, the archaeologist has quite simply allowed his political and cultural sympathies to cloud his reasoning. But there is no justification for such intellectual confusion: we can perfectly well remember the victims of a horrible genocide, and support their descendants' valid political goals, without endorsing uncritically (or hypocritically) their societies' traditional creation myths. (After all, if you want to support Native American land claims, does it *really* matter whether Native Americans have been in North America "forever" or *merely* for 10,000 years?) Moreover, the relativists' stance is extremely condescending: it treats a complex society as a monolith, obscures the conflicts within it, and takes its most obscurantist factions as spokespeople for the whole.

5. *The traditional philosophical and literary training.* We have no desire to criticize this training as such; indeed, it is probably adequate for the goals it pursues. Nevertheless, it may be a handicap when one turns to scientific texts, for two reasons.

First of all, the author or the literality of the text have, in literature or even in philosophy, a relevance they do not have in science. One can learn physics without ever reading Galileo, Newton, or Einstein, and study biology without reading a line of Darwin.[260] What matters are the factual and theoretical arguments these authors offer, not the words they used. Besides,

---

[259]When challenged, relativist anthropologists sometimes *deny* that there is a distinction between knowledge (i.e. justified true belief) and mere belief, by denying that beliefs—even cognitive beliefs about the external world—can be objectively (trans-culturally) true or false. But it is hard to take such a claim seriously. Didn't millions of Native Americans *really* die in the period following the European invasion? Is this merely a belief held to be true within some cultures?

[260]Which is not to say that the student or the researcher cannot *profit* from reading classical texts. It all depends upon the pedagogical qualities of the authors in question. For example, physicists today can read Galileo and Einstein both for the sheer pleasure of their writing and for their deep insight. And biologists can certainly do likewise with Darwin.

their ideas may have been radically modified or even overturned by subsequent developments in their disciplines. Furthermore, scientists' personal qualities and extra-scientific beliefs are irrelevant to the evaluation of their theories. Newton's mysticism and alchemy, for example, are important for the history of science and more generally for the history of human thought, but not for physics.

The second problem comes from the privilege granted to theories over experiments (which is related to the privilege granted to texts over facts). The link between a scientific theory and its experimental test is often extremely complex and indirect. Therefore, a philosopher will tend to approach the sciences preferentially through their conceptual aspect (so do we, in fact). But the whole problem comes precisely from the fact that, if one does not *also* take into account the empirical aspects, then scientific discourse indeed becomes nothing more than a "myth" or "narration" among many others.

The Role of Politics

> It's not we who lord it over things, it seems, but things which lord it over us. But that's only because some people make use of things in order to lord it over others. We shall only be freed from the forces of nature when we are free of human force. Our knowledge of nature must be supplemented with a knowledge of human society if we are to use our knowledge of nature in a human way.
> —*Bertolt Brecht (1965 [1939–1940], pp. 42–43)*

The origins of postmodernism are not purely intellectual. Both philosophical relativism and the works of the authors analyzed here have had a specific appeal to some political tendencies that can be characterized (or characterize themselves) as left-wing or progressive. Moreover, the "science wars" are often viewed as a political conflict between "progressives" and "conservatives".[261] Of course, there is also a long anti-rationalist tradition in some right-wing movements, but what is new and curious about postmodernism is that it is an anti-rationalist form of thought that has seduced part of the left.[262] We shall try here to analyze how this sociological link came to be, and to explain why it is due, in our opinion, to a number of conceptual confusions. We shall limit ourselves mainly to the situation in the United States, where the link between postmodernism and some tendencies on the political left is particularly clear.

When one discusses a set of ideas, such as postmodernism, from a political point of view, it is important to distinguish carefully between the intrinsic intellectual value of those ideas, the objective political role they play, and the subjective reasons for which various people defend or attack them. Now, it often happens that a given social group shares two ideas (or two groups of ideas), call them A and B. Let us suppose that A is relatively valid, that B is much less valid, and that there is no real logical link between the two. People belonging to the social group will often try to legitimize B by invoking the validity of A and the existence of a sociological link between A and B. Conversely, their opponents will try to denigrate A by citing the lack of validity of B and the same sociological link.[263]

The existence of such a link between postmodernism and the left constitutes, prima facie, a serious paradox. For most of the past two centuries, the left has been identified with science and against obscurantism, believing that rational thought and the fearless analysis of objective reality (both natural and social) are incisive tools for combating the mystifications promoted by the powerful—not to mention being desirable human ends in their own right. And yet, over the past two decades, a large number of "progressive" or "leftist" academic humanists

---

[261]Extreme versions of this idea can be found, for example, in Ross (1995) and Harding (1996).

[262]But not only the left: see the quotation from Václav Havel on p. 192 above.

[263]A similar observation holds when a famous individual holds ideas of type A and B.

and social scientists (though virtually no natural scientists, whatever their political views) have turned away from this Enlightenment legacy and—bolstered by French imports such as deconstruction as well as by home-grown doctrines like feminist standpoint epistemology—have embraced one or another version of epistemic relativism. Our aim here is to understand the causes of this historical *volte-face*.

We shall distinguish three types of intellectual sources linked to the emergence of postmodernism within the political left[264]:

1. *The new social movements.* The 1960s and 1970s saw the rise of new social movements—the black liberation movement, the feminist movement, and the gay-rights movement, among others—struggling against forms of oppression that had largely been underestimated by the traditional political left. More recently, some tendencies within these movements have concluded that postmodernism, in one form or another, is the philosophy most suited to their aspirations.

There are two separate issues to discuss. One is conceptual: is there a logical link, in either direction, between the new social movements and postmodernism? The other is sociological: to what extent have the members of these movements embraced postmodernism, and for what reasons?

One factor driving the new social movements toward postmodernism was, undoubtedly, a dissatisfaction with the old leftist orthodoxies. The traditional left, in both its Marxist and non-Marxist variants, generally saw itself as the rightful inheritor of the Enlightenment and as the embodiment of science and rationality. Moreover, Marxism explicitly linked philosophical materialism to a theory of history giving primacy—in some versions, near-exclusivity—to economic and class struggles. The evident narrowness of this latter perspective understandably led some currents within the new social movements to reject, or at least distrust, science and rationality as such.

---

[264]For a more detailed discussion, see Eagleton (1995) and Epstein (1995, 1997).

But this is a conceptual error, which mirrors an identical error committed by the traditional Marxist left. In fact, concrete socio-political theories can never be deduced logically from abstract philosophical schemes; and conversely, there is no unique philosophical position compatible with a given socio-political program. In particular, as Bertrand Russell observed long ago, there is no logical connection between philosophical materialism and Marxian historical materialism. Philosophical materialism is compatible with the idea that history is determined primarily by religion, sexuality, or climate (which would run counter to historical materialism); and conversely, economic factors could be the primary determinants of human history even if mental events were sufficiently independent of physical events to make philosophical materialism false. Russell concludes: "It is of some moment to realize such facts as this, because otherwise political theories are both supported and opposed for quite irrelevant reasons, and arguments of theoretical philosophy are employed to determine questions which depend upon concrete facts of human nature. This mixture damages both philosophy and politics, and is therefore important to avoid."[265]

The sociological link between postmodernism and the new social movements is an exceedingly complicated one. A satisfactory analysis would require, at the very least, disentangling the various strands that compose "postmodernism" (as the logical relations between them are quite weak), treating each of the new social movements individually (as their histories are quite different), sorting out the distinct currents within these movements, and distinguishing the roles played by activists and theorists. This is a problem requiring (dare we say it?) careful empirical investigation, and we leave it to sociologists and intellectual historians. Let us nevertheless state our *conjecture* that the new social movements' penchant for postmodernism exists mostly within academia and is much weaker than both

---

[265]Russell (1949 [1920], p. 80), reprinted in Russell (1961b, pp. 528–529).

the postmodernist left and the traditionalist right generally portray it to be.[266]

2. *Political discouragement.* Another source of postmodern ideas is the desperate situation and general disorientation of the left, a situation that appears to be unique in its history. The communist regimes have collapsed; the social-democratic parties, where they remain in power, apply watered-down neoliberal policies; and the Third World movements that led their countries to independence have, in most cases, abandoned any attempt at autonomous development. In short, the harshest form of "free market" capitalism seems to have become the implacable reality for the foreseeable future. Never before have the ideals of justice and equality seemed so utopian. Without entering into an analysis of the causes of this situation (much less proposing solutions), it is easy to understand that it generates a kind of discouragement that expresses itself in part in postmodernism. The linguist and activist Noam Chomsky has described this evolution very well[267]:

> If you really feel, Look, it's too hard to deal with real problems, there are lots of ways to avoid doing so. One of them is to go off on wild goose chases that don't matter. Another is to get involved in academic cults that are very divorced from any reality and that provide a defense against dealing with the world as it actually is. There's plenty of that going on, including in the left. I just saw some very depressing examples of it in my trip to Egypt a couple of weeks ago. I was there to talk on international affairs. There's a very lively, civilized intellectual community, very courageous people who spent years in Nasser's jails being practically tortured to death and came out struggling. Now throughout the Third World there's a sense of great despair and hopelessness. The way it showed up there, in very educated circles with European connections, was to become totally immersed in the latest lunacies of Paris

---

[266]For further analysis, see Epstein (1995, 1997).

[267]See also Eagleton (1995).

culture and to focus totally on those. For example, when I
would give talks about current realities, even in research in-
stitutes dealing with strategic issues, participants wanted it to
be translated into post-modern gibberish. For example, rather
than have me talk about the details of what's going on in U.S.
policy or the Middle East, where they live, which is too grubby
and uninteresting, they would like to know how does modern
linguistics provide a new paradigm for discourse about international affairs that will supplant the post-structuralist text.
That would really fascinate them. But not what do Israeli cab-
inet records show about internal planning. That's really depressing. (Chomsky 1994, pp. 163–164)

In this way, the remnants of the left have collaborated in driving
the last nail in the coffin of the ideals of justice and progress. We
modestly suggest letting in a little bit of air, in the hope that one
day the corpse will awaken.

3. *Science as an easy target.* In this atmosphere of general
discouragement, it is tempting to attack something that is sufficiently linked to the powers-that-be so as not to appear very
sympathetic, but sufficiently weak to be a more-or-less accessible target (since the concentration of power and money are beyond reach). Science fulfills these conditions, and this partly
explains the attacks against it. In order to analyze these attacks,
it is crucial to distinguish at least four different senses of the
word "science": an intellectual endeavor aimed at a rational understanding of the world; a collection of accepted theoretical
and experimental ideas; a social community with particular
mores, institutions and links to the larger society; and, finally,
applied science and technology (with which science is often
confused). All too frequently, valid critiques of "science", understood in one of these senses, are taken to be arguments
against science in a different sense.[268] Thus, it is undeniable that

---

[268]For an example of such confusions, see the essay of Raskin and Bernstein (1987,
pp. 69–103); and for a good dissection of these confusions, see the responses by
Chomsky in the same volume (pp. 104–156).

science, as a social institution, is linked to political, economic, and military power, and that the social role played by scientists is often pernicious. It is also true that technology has mixed results—sometimes disastrous ones—and that it rarely yields the miracle solutions that its most fervent advocates regularly promise.[269] Finally, science, considered as a body of knowledge, is always fallible, and scientists' errors are sometimes due to all sorts of social, political, philosophical, or religious prejudices. We are in favor of reasonable criticisms of science understood in all these senses. In particular, the critiques of science viewed as a body of knowledge—at least those that are most convincing—follow, in general, a standard pattern: First one shows, using conventional scientific arguments, why the research in question is flawed according to the ordinary canons of good science; then, and only then, one attempts to explain how the researchers' social prejudices (which may well have been unconscious) led them to violate these canons. One may be tempted to jump directly to the second step, but the critique then loses much of its force.

Unfortunately, some critiques go beyond attacking the worst aspects of science (militarism, sexism, etc.) and attack its best aspects: the attempt at rationally understanding the world, and the scientific method, understood broadly as a respect for empirical evidence and for logic.[270] It is naive to believe that it is not the rational attitude itself that is really challenged by postmodernism. Moreover, this aspect is an easy target, because any attack on rationality can find a host of allies: all those who believe in superstitions, be they traditional ones (e.g. religious fundamentalism) or New Age.[271] If one adds to that a facile confusion

---

[269]It must nevertheless be emphasized that technology is often blamed for consequences that are due more to the social structure than to technology itself.

[270]Let us note, in passing, that it is precisely the emphasis on objectivity and verification that offers the best protection against ideological bias masquerading as science.

[271]According to recent polls, 47% of Americans believe in the creation account of Genesis, 49% in possession by the devil, 36% in telepathy, and 25% in astrology. Mercifully, only 11% believe in channeling, and 7% in the healing power of

between science and technology, one arrives at a struggle that is relatively popular, though not particularly progressive.

Those who wield political or economic power will quite naturally prefer that science and technology be attacked as such, because these attacks help conceal the relationships of force on which their own power is based. Furthermore, by attacking rationality, the postmodern left deprives itself of a powerful instrument for criticizing the existing social order. Chomsky observes that, in a not-so-distant past,

> Left intellectuals took an active part in the lively working class culture. Some sought to compensate for the class character of the cultural institutions through programs of workers' education, or by writing best-selling books on mathematics, science, and other topics for the general public. Remarkably, their left counterparts today often seek to deprive working people of these tools of emancipation, informing us that the "project of the Enlightenment" is dead, that we must abandon the "illusions" of science and rationality—a message that will gladden the hearts of the powerful, delighted to monopolize these instruments for their own use. (Chomsky 1993, p. 286)

Finally, let us briefly discuss the subjective motivations of those who are opposed to postmodernism. These are complicated to analyze, and the reactions that followed the publication of Sokal's parody suggest a prudent reflection. Many people are simply irritated by the arrogance and empty verbiage of postmodernist discourse and by the spectacle of an intellectual community where everyone repeats sentences that no one understands. It goes without saying that we share, with some nuances, this attitude.

But other reactions are much less pleasant, and they are good illustrations of the confusion between sociological and

---

pyramids. For detailed data and references to the original sources, see Sokal (1996c, note 17), reprinted here in Appendix C.

logical links. For example, the *New York Times* presented the "Sokal affair" as a debate between conservatives who believe in objectivity, at least as a goal, and leftists who deny it. Obviously, the situation is more complex. Not all those on the political left reject the goal (however imperfectly realized) of objectivity[272]; and there is not, in any case, a simple logical relation between political and epistemological views.[273] Other commentators link this story to attacks against "multiculturalism" and "political correctness". It would take us much too far afield to discuss these questions in detail, but let us emphasize that we in no way reject the openness to other cultures or the respect for minorities that are often ridiculed in these kinds of attacks.

Why Does It Matter?

> The concept of "truth" as something dependent upon facts largely outside human control has been one of the ways in which philosophy hitherto has inculcated the necessary element of humility. When this check upon pride is removed, a further step is taken on the road towards a certain kind of madness—the intoxication of power which invaded philosophy with Fichte, and to which modern men, whether philosophers or not, are prone. I am persuaded that this intoxication is the greatest danger of our time, and that any philosophy which, however unintentionally, contributes to it is increasing the danger of vast social disaster.
> —*Bertrand Russell, History of Western Philosophy*
> *(1961a, p. 782)*

Why spend so much time exposing these abuses? Do the postmodernists represent a real danger? Certainly not for the natural sciences, at least not at present. The problems faced today by the natural sciences concern primarily the financing of

---

[272]See, for example, Chomsky (1992–93), Ehrenreich (1992–93), Albert (1992–93, 1996), and Epstein (1997) among many others.

[273]Much further down in the *New York Times* article (Scott 1996), the reporter mentions Sokal's leftist political positions and the fact that he taught mathematics in Nicaragua during the Sandinista government. But the contradiction is not even noticed, much less resolved.

research, and in particular the threat posed to scientific objectivity when public funding is increasingly replaced by private sponsorship. But postmodernism has nothing to do with this.[274] It is, rather, the *social* sciences that suffer when fashionable nonsense and word games displace the critical and rigorous analysis of social realities.

Postmodernism has three principal negative effects: a waste of time in the human sciences, a cultural confusion that favors obscurantism, and a weakening of the political left.

First of all, postmodern discourse, exemplified by the texts we quote, functions in part as a dead end in which some sectors of the humanities and social sciences have gotten lost. No research, whether on the natural or the social world, can progress on a basis that is both conceptually confused and radically detached from empirical evidence.

It could be argued that the authors of the texts quoted here have no real impact on research because their lack of professionalism is well-known in academic circles. This is only partly true: it depends on the authors, the countries, the fields of study, and the eras. For example, the works of Barnes–Bloor and Latour have had an undeniable influence in the sociology of science, even if they have never been hegemonic. The same holds true for Lacan and Deleuze–Guattari in certain areas of literary theory and cultural studies, and for Irigaray in women's studies.

What is worse, in our opinion, is the adverse effect that abandoning clear thinking and clear writing has on teaching and culture. Students learn to repeat and to embellish discourses that they only barely understand. They can even, if they are lucky, make an academic career out of it by becoming expert in the manipulation of an erudite jargon.[275] After all, one of us managed, after only three months of study, to master the postmod-

---

[274]Note, however, that postmodernists and relativists are ill-placed to *criticize* this threat to scientific objectivity, since they deny objectivity even as a goal.

[275]This phenomenon is by no means due to postmodernism—Andreski (1972) illustrated it brilliantly for the traditional social sciences—and it is also present, to a much lesser extent, in the natural sciences. Nevertheless, the obscurity of

ern lingo well enough to publish an article in a prestigious journal. As commentator Katha Pollitt astutely noted, "the comedy of the Sokal incident is that it suggests that even the postmodernists don't really understand one another's writing and make their way through the text by moving from one familiar name or notion to the next like a frog jumping across a murky pond by way of lily pads."[276] The deliberately obscure discourses of postmodernism, and the intellectual dishonesty they engender, poison a part of intellectual life and strengthen the facile anti-intellectualism that is already all too widespread in the general public.

The lackadaisical attitude toward scientific rigor that one finds in Lacan, Kristeva, Baudrillard, and Deleuze had an undeniable success in France during the 1970s and is still remarkably influential there.[277] This way of thinking spread outside France, notably in the English-speaking world, during the 1980s and 1990s. Conversely, cognitive relativism developed during the 1970s mostly in the English-speaking world (for example, with the beginning of the "strong programme") and spread later to France.

These two attitudes are, of course, conceptually distinct; one can be adopted with or without the other. However, they are indirectly linked: if anything, or almost anything, can be read into the content of scientific discourse, then why should anyone take science seriously as an objective account of the world? Conversely, if one adopts a relativist philosophy, then arbitrary comments on scientific theories become legitimate. Relativism and sloppiness are therefore mutually reinforcing.

But the most serious cultural consequences of relativism come from its application to the social sciences. The British historian Eric Hobsbawm has eloquently decried

postmodernist jargon, and its almost total lack of contact with concrete realities, exacerbate this situation.

[276]Pollitt (1996).

[277]In the French edition we wrote "but is undoubtedly somewhat passé there", but contacts we have had since the publication of our book have led us to rethink. For example, Lacanianism is extraordinarily influential in French psychiatry.

the rise of "postmodernist" intellectual fashions in Western universities, particularly in departments of literature and anthropology, which imply that all "facts" claiming objective existence are simply intellectual constructions. In short, that there is no clear difference between fact and fiction. But there is, and for historians, even for the most militantly antipositivist ones among us, the ability to distinguish between the two is absolutely fundamental. (Hobsbawm 1993, p. 63)

Hobsbawm goes on to show how rigorous historical work can refute the fictions propounded by reactionary nationalists in India, Israel, the Balkans, and elsewhere, and how the postmodernist attitude disarms us in the face of these threats.

At a time when superstitions, obscurantism, and nationalist and religious fanaticism are spreading in many parts of the world—including the "developed" West—it is irresponsible, to say the least, to treat with such casualness what has historically been the principal defense against these follies, namely a rational vision of the world. It is doubtless not the intention of postmodernist authors to favor obscurantism, but it is an inevitable consequence of their approach.

Finally, for all those of us who identify with the political left, postmodernism has specific negative consequences. First of all, the extreme focus on language and the elitism linked to the use of a pretentious jargon contribute to enclosing intellectuals in sterile debates and to isolating them from social movements taking place outside their ivory tower. When progressive students arriving on American campuses learn that the most radical idea (even politically) is to adopt a thoroughly skeptical attitude and to immerse oneself completely in textual analysis, their energy—which could be fruitfully employed in research and organizing—is squandered. Second, the persistence of confused ideas and obscure discourses in some parts of the left tends to discredit the entire left; and the right does not pass up the opportunity to exploit this connection demagogically.[278]

---

[278]See, for example, Kimball (1990) and D'Souza (1991).

But the most important problem is that any possibility of a social critique that could reach those who are not already convinced—a necessity, given the present infinitesimal size of the American left—becomes logically impossible, due to the subjectivist presuppositions.[279] If all discourses are merely "stories" or "narrations", and none is more objective or truthful than another, then one must concede that the worst sexist or racist prejudices and the most reactionary socio-economic theories are "equally valid", at least as descriptions or analyses of the real world (assuming that one admits the existence of a real world). Clearly, relativism is an extremely weak foundation on which to build a criticism of the existing social order.

If intellectuals, particularly those on the left, wish to make a positive contribution to the evolution of society, they can do so above all by clarifying the prevailing ideas and by demystifying the dominant discourses, not by adding their own mystifications. A mode of thought does not become "critical" simply by attributing that label to itself, but by virtue of its content.

To be sure, intellectuals tend to exaggerate their impact on the larger culture, and we want to avoid falling into this trap. We think, nevertheless, that the ideas—even the most abstruse ones—taught and debated within universities have, over time, cultural effects beyond academia. Bertrand Russell undoubtedly exaggerated when he denounced the perverse social consequences of confusion and subjectivism, but his fears were not entirely unfounded.

What Next?

"A spectre is haunting U.S. intellectual life: the spectre of Left Conservatism." So proclaimed the announcement for a recent

---

[279] The word "logically" is important here. In practice, some individuals use postmodern language while opposing racist or sexist discourses with perfectly rational arguments. We think, simply, that there is an incoherence here between their practice and their avowed philosophy (which may not be such a horrible thing).

conference at the University of California–Santa Cruz, where
we and others[280] were criticized for our opposition to "anti-
foundationalist [i.e. postmodernist] theoretical work" and—
horror of horrors—for "an attempt at consensus-building . . .
founded on notions of the real". We were portrayed as socially
conservative Marxists trying to marginalize feminist, gay, and
racial-justice politics, and as sharing the values of Rush Limbaugh.[281] Might these lurid accusations symbolize, albeit in an
extreme way, what has gone wrong with postmodernism?

Throughout this book, we have defended the idea that there
is such a thing as evidence and that facts matter. However,
many questions of vital interest—notably those concerning the
future—cannot be answered conclusively on the basis of evidence and reason, and they lead human beings to indulge in
(more-or-less-informed) speculation. We would like to end this
book with a bit of speculation of our own, concerning the future
of postmodernism. As we have repeatedly stressed, postmodernism is such a complicated network of ideas—with only
weak logical links between them—that it is difficult to characterize it more precisely than as a vague zeitgeist. Nevertheless,
the roots of this zeitgeist are not hard to identify, and go back
to the early 1960s: challenges to empiricist philosophies of science with Kuhn, critiques of humanist philosophies of history
with Foucault, disillusionment with grand schemes for political
change. Like all new intellectual currents, postmodernism, in its
inchoate phase, met with resistance from the old guard. But
new ideas have the privilege of youth playing for them, and the
resistance turned out to be vain.

Almost forty years later, revolutionaries have aged and marginality has become institutionalized. Ideas that contained some
truth, if properly understood, have degenerated into a vulgate
that mixes bizarre confusions with overblown banalities. It

---

[280]Notably the feminist writers Barbara Ehrenreich and Katha Pollitt and the leftist
filmmaker Michael Moore.

[281]Accounts of the Left Conservatism conference can be found in Sand (1998), Willis
et al. (1998), Dumm *et al.* (1998), and Zarlengo (1998).

seems to us that postmodernism, whatever usefulness it originally had as a corrective to hardened orthodoxies, has lived this out and is now running its natural course. Although the name was not ideally chosen to invite a succession (what can come after *post-*?), we are under the inescapable impression that times are changing. One sign is that the challenge comes nowadays not only from the rearguard, but also from people who are neither die-hard positivists nor old-fashioned Marxists, and who understand the problems encountered by science, rationality, and traditional leftist politics—but who believe that criticism of the past should enlighten the future, not lead to contemplation of the ashes.[282]

What will come after postmodernism? Since the principal lesson to be learned from the past is that predicting the future is hazardous, we can only list our fears and our hopes. One possibility is a backlash leading to some form of dogmatism, mysticism (e.g. New Age), or religious fundamentalism. This may appear unlikely, at least in academic circles, but the demise of reason has been radical enough to pave the way for a more extreme irrationalism. In this case intellectual life would go from bad to worse. A second possibility is that intellectuals will become reluctant (at least for a decade or two) to attempt any thoroughgoing critique of the existing social order, and will either become its servile advocates—as some formerly leftist French intellectuals did after 1968—or retreat from political engagement entirely. Our hopes, however, go in a different direction: the emergence of an intellectual culture that would be rationalist but not dogmatic, scientifically minded but not scientistic, open-minded but not frivolous, and politically progressive but not sectarian. But this, of course, is only a hope, and perhaps only a dream.

---

[282]Another encouraging sign is that some of the most insightful commentary has been produced by students, both in France (Coutty 1998) and in the U.S. (Sand 1998).

A. Transgressing the Boundaries: Toward a Transformative Hermeneutics of Quantum Gravity*

> Transgressing disciplinary boundaries . . . [is] a subversive
> undertaking since it is likely to violate the sanctuaries of accepted
> ways of perceiving. Among the most fortified boundaries have
> been those between the natural sciences and the humanities.
> —*Valerie Greenberg, Transgressive Readings (1990, p. 1)*

> The struggle for the transformation of ideology into critical
> science . . . proceeds on the foundation that the critique of all
> presuppositions of science and ideology must be the only absolute
> principle of science.
> —*Stanley Aronowitz, Science as Power (1988b, p. 339)*

There are many natural scientists, and especially physicists, who continue to reject the notion that the disciplines concerned with social and cultural criticism can have anything to contribute, except perhaps peripherally, to their research. Still less are they receptive to the idea that the very foundations of their worldview must be revised or rebuilt in the light of such criticism. Rather, they cling to the dogma imposed by the long post-Enlightenment hegemony over the Western intellectual outlook,

*Originally published in *Social Text* **#46/47** (spring/summer 1996), pp. 217–252. © Duke University Press.

which can be summarized briefly as follows: that there exists an external world, whose properties are independent of any individual human being and indeed of humanity as a whole; that these properties are encoded in "eternal" physical laws; and that human beings can obtain reliable, albeit imperfect and tentative, knowledge of these laws by hewing to the "objective" procedures and epistemological strictures prescribed by the (so-called) scientific method.

But deep conceptual shifts within twentieth-century science have undermined this Cartesian-Newtonian metaphysics[1]; revisionist studies in the history and philosophy of science have cast further doubt on its credibility[2]; and, most recently, feminist and poststructuralist critiques have demystified the substantive content of mainstream Western scientific practice, revealing the ideology of domination concealed behind the façade of "objectivity".[3] It has thus become increasingly apparent that physical "reality", no less than social "reality", is at bottom a social and linguistic construct; that scientific "knowledge", far from being objective, reflects and encodes the dominant ideologies and power relations of the culture that produced it; that the truth claims of science are inherently theory-laden and self-referential; and consequently, that the discourse of the scientific community, for all its undeniable value, cannot assert a privileged epistemological status with respect to counter-hegemonic narratives emanating from dissident or marginalized communities. These themes can be traced, despite some differences of emphasis, in Aronowitz's analysis of the cultural fabric that produced quantum mechanics[4]; in Ross' discussion of oppositional discourses in post-quantum science[5]; in Irigaray's and Hayles'

---

[1]Heisenberg (1958), Bohr (1963).

[2]Kuhn (1970), Feyerabend (1975), Latour (1987), Aronowitz (1988b), Bloor (1991).

[3]Merchant (1980), Keller (1985), Harding (1986, 1991), Haraway (1989, 1991), Best (1991).

[4]Aronowitz (1988b, especially chaps. 9 and 12).

[5]Ross (1991, introduction and chap. 1).

exegeses of gender encoding in fluid mechanics[6]; and in Harding's comprehensive critique of the gender ideology underlying the natural sciences in general and physics in particular.[7]

Here my aim is to carry these deep analyses one step farther, by taking account of recent developments in quantum gravity: the emerging branch of physics in which Heisenberg's quantum mechanics and Einstein's general relativity are at once synthesized and superseded. In quantum gravity, as we shall see, the space-time manifold ceases to exist as an objective physical reality; geometry becomes relational and contextual; and the foundational conceptual categories of prior science— among them, existence itself—become problematized and relativized. This conceptual revolution, I will argue, has profound implications for the content of a future postmodern and liberatory science.

My approach will be as follows: First I will review very briefly some of the philosophical and ideological issues raised by quantum mechanics and by classical general relativity. Next I will sketch the outlines of the emerging theory of quantum gravity, and discuss some of the conceptual issues it raises. Finally, I will comment on the cultural and political implications of these scientific developments. It should be emphasized that this article is of necessity tentative and preliminary; I do not pretend to answer all of the questions that I raise. My aim is, rather, to draw the attention of readers to these important developments in physical science, and to sketch as best I can their philosophical and political implications. I have endeavored here to keep mathematics to a bare minimum; but I have taken care to provide references where interested readers can find all requisite details.

---

[6]Irigaray (1985), Hayles (1992).

[7]Harding (1986, especially chaps. 2 and 10); Harding (1991, especially chap. 4).

Quantum Mechanics:
Uncertainty, Complementarity, Discontinuity, and
Interconnectedness

It is not my intention to enter here into the extensive debate on the conceptual foundations of quantum mechanics.[8] Suffice it to say that anyone who has seriously studied the equations of quantum mechanics will assent to Heisenberg's measured (pardon the pun) summary of his celebrated *uncertainty principle*:

> We can no longer speak of the behaviour of the particle independently of the process of observation. As a final consequence, the natural laws formulated mathematically in quantum theory no longer deal with the elementary particles themselves but with our knowledge of them. Nor is it any longer possible to ask whether or not these particles exist in space and time objectively . . .
>
> When we speak of the picture of nature in the exact science of our age, we do not mean a picture of nature so much as a *picture of our relationships with nature*. . . . Science no longer confronts nature as an objective observer, but sees itself as an actor in this interplay between man [*sic*] and nature. The scientific method of analysing, explaining and classifying has become conscious of its limitations, which arise out of the fact that by its intervention science alters and refashions the object of investigation. In other words, method and object can no longer be separated.[9,10]

---

[8]For a sampling of views, see Jammer (1974), Bell (1987), Albert (1992), Dürr, Goldstein and Zanghí (1992), Weinberg (1992, chap. IV), Coleman (1993), Maudlin (1994), Bricmont (1994).

[9]Heisenberg (1958, pp. 15, 28–29), emphasis in Heisenberg's original. See also Overstreet (1980), Craige (1982), Hayles (1984), Greenberg (1990), Booker (1990), and Porter (1990) for examples of cross-fertilization of ideas between relativistic quantum theory and literary criticism.

[10]Unfortunately, Heisenberg's uncertainty principle has frequently been misinterpreted by amateur philosophers. As Gilles Deleuze and Félix Guattari (1994, pp. 129–130) lucidly point out,

Along the same lines, Niels Bohr wrote:

> An independent reality in the ordinary physical sense can . . .
> neither be ascribed to the phenomena nor to the agencies of
> observation.[11]

Stanley Aronowitz has convincingly traced this worldview to
the crisis of liberal hegemony in Central Europe in the years
prior and subsequent to World War I.[12,13]

A second important aspect of quantum mechanics is its principle of *complementarity* or *dialecticism*. Is light a particle or
a wave? Complementarity "is the realization that particle and
wave behavior are mutually exclusive, yet that both are necessary for a complete description of all phenomena."[14] More generally, notes Heisenberg,

---

in quantum physics, Heisenberg's demon does not express the impossibility of
measuring both the speed and the position of a particle on the grounds of a
subjective interference of the measure with the measured, but it measures
exactly an objective state of affairs that leaves the respective position of two of
its particles outside of the field of its actualization, the number of independent
variables being reduced and the values of the coordinates having the same
probability . . . . Perspectivism, or scientific relativism, is never relative to a
subject: it constitutes not a relativity of truth but, on the contrary, a truth of the
relative, that is to say, of variables whose cases it orders according to the values
it extracts from them in its system of coordinates . . .

[11]Bohr (1928), cited in Pais (1991, p. 314).

[12]Aronowitz (1988b, pp. 251–256).

[13]See also Porush (1989) for a fascinating account of how a second group of scientists
and engineers—cyberneticists—contrived, with considerable success, to subvert the
most revolutionary implications of quantum physics. The main limitation of Porush's
critique is that it remains solely on a cultural and philosophical plane; his conclusions
would be immeasurably strengthened by an analysis of economic and political
factors. (For example, Porush fails to mention that engineer-cyberneticist Claude
Shannon worked for the then–telephone monopoly AT&T.) A careful analysis would
show, I think, that the victory of cybernetics over quantum physics in the 1940s and
50s can be explained in large part by the centrality of cybernetics to the ongoing
capitalist drive for automation of industrial production, compared to the marginal
industrial relevance of quantum mechanics.

[14]Pais (1991, p. 23). Aronowitz (1981, p. 28) has noted that wave-particle duality
renders the "will to totality in modern science" severely problematic:

> The differences within physics between wave and particle theories of matter, the
> indeterminacy principle discovered by Heisenberg, Einstein's relativity theory,
> all are accommodations to the impossibility of arriving at a unified field theory,
> one in which the "anomaly" of difference for a theory which posits identity may
> be resolved without challenging the presuppositions of science itself.

the different intuitive pictures which we use to describe atomic systems, although fully adequate for given experiments, are nevertheless mutually exclusive. Thus, for instance, the Bohr atom can be described as a small-scale planetary system, having a central atomic nucleus about which the external electrons revolve. For other experiments, however, it might be more convenient to imagine that the atomic nucleus is surrounded by a system of stationary waves whose frequency is characteristic of the radiation emanating from the atom. Finally, we can consider the atom chemically. . . . Each picture is legitimate when used in the right place, but the different pictures are contradictory and therefore we call them mutually complementary.[15]

And once again Bohr:

A complete elucidation of one and the same object may require diverse points of view which defy a unique description. Indeed, strictly speaking, the conscious analysis of any concept stands in a relation of exclusion to its immediate application.[16]

For further development of these ideas, see Aronowitz (1988a, pp. 524–525, 533).

[15]Heisenberg (1958, pp. 40–41).

[16]Bohr (1934), cited in Jammer (1974, p. 102). Bohr's analysis of the complementarity principle also led him to a social outlook which was, for its time and place, notably progressive. Consider the following excerpt from a 1938 lecture (Bohr 1958, p. 30):

> I may perhaps here remind you of the extent to which in certain societies the roles of men and women are reversed, not only regarding domestic and social duties but also regarding behaviour and mentality. Even if many of us, in such a situation, might perhaps at first shrink from admitting the possibility that it is entirely a caprice of fate that the people concerned have their specific culture and not ours, and we not theirs instead of our own, it is clear that even the slightest suspicion in this respect implies a betrayal of the national complacency inherent in any human culture resting in itself.

This foreshadowing of postmodernist epistemology is by no means coincidental. The profound connections between complementarity and deconstruction have recently been elucidated by Froula[17] and Honner[18], and, in great depth, by Plotnitsky.[19,20,21]

A third aspect of quantum physics is *discontinuity* or *rupture*: as Bohr explained,

> [the] essence [of the quantum theory] may be expressed in the so-called quantum postulate, which attributes to any atomic process an essential discontinuity, or rather individuality, completely foreign to the classical theories and symbolized by Planck's quantum of action.[22]

---

[17]Froula (1985).

[18]Honner (1994).

[19]Plotnitsky (1994). This impressive work also explains the intimate connections with Gödel's proof of the incompleteness of formal systems and with Skolem's construction of nonstandard models of arithmetic, as well as with Bataille's general economy. For further discussion of Bataille's physics, see Hochroth (1995).

[20]Numerous other examples could be adduced. For instance, Barbara Johnson (1989, p. 12) makes no specific reference to quantum physics; but her description of deconstruction is an eerily exact summary of the complementarity principle:

> Instead of a simple "either/or" structure, deconstruction attempts to elaborate a discourse that says *neither* "either/or", *nor* "both/and" nor even "neither/nor", while at the same time not totally abandoning these logics either.

See also McCarthy (1992) for a thought-provoking analysis that raises disturbing questions about the "complicity" between (nonrelativistic) quantum physics and deconstruction.

[21]Permit me in this regard a personal recollection: Fifteen years ago, when I was a graduate student, my research in relativistic quantum field theory led me to an approach which I called "de[con]structive quantum field theory" (Sokal 1982). Of course, at that time I was completely ignorant of Jacques Derrida's work on deconstruction in philosophy and literary theory. In retrospect, however, there is a striking affinity: my work can be read as an exploration of how the orthodox discourse (e.g. Itzykson and Zuber 1980) on scalar quantum field theory in four-dimensional space-time (in technical terms, "renormalized perturbation theory" for the φ_4^4 theory) can be seen to assert its own unreliability and thereby to undermine its own affirmations. Since then, my work has shifted to other questions, mostly connected with phase transitions; but subtle homologies between the two fields can be discerned, notably the theme of discontinuity (see Notes 22 and 81 below). For further examples of deconstruction in quantum field theory, see Merz and Knorr Cetina (1994).

[22]Bohr (1928), cited in Jammer (1974, p. 90).

A half-century later, the expression "quantum leap" has so entered our everyday vocabulary that we are likely to use it without any consciousness of its origins in physical theory.

Finally, Bell's theorem[23] and its recent generalizations[24] show that an act of observation here and now can affect not only the object being observed—as Heisenberg told us—but also an object *arbitrarily far away* (say, on Andromeda galaxy). This phenomenon—which Einstein termed "spooky"—imposes a radical reevaluation of the traditional mechanistic concepts of space, object, and causality[25], and suggests an alternative worldview in which the universe is characterized by interconnectedness and (w)holism: what physicist David Bohm has called "implicate order".[26] New Age interpretations of these insights from quantum physics have often gone overboard in unwarranted speculation, but the general soundness of the ar-

---

[23]Bell (1987, especially chaps. 10 and 16). See also Maudlin (1994, chap. 1) for a clear account presupposing no specialized knowledge beyond high-school algebra.

[24]Greenberger *et al.* (1989, 1990), Mermin (1990, 1993).

[25]Aronowitz (1988b, p. 331) has made a provocative observation concerning nonlinear causality in quantum mechanics and its relation to the social construction of time:

> Linear causality assumes that the relation of cause and effect can be expressed as a function of temporal succession. Owing to recent developments in quantum mechanics, we can postulate that it is possible to know the effects of absent causes; that is, speaking metaphorically, effects may anticipate causes so that our perception of them may precede the physical occurrence of a "cause." The hypothesis that challenges our conventional conception of linear time and causality and that asserts the possibility of time's reversal also raises the question of the degree to which the concept of "time's arrow" is inherent in all scientific theory. If these experiments are successful, the conclusions about the way time as "clock-time" has been constituted historically will be open to question. We will have "proved" by means of experiment what has long been suspected by philosophers, literary and social critics: that time is, in part, a conventional construction, its segmentation into hours and minutes a product of the need for industrial discipline, for rational organization of social labor in the early bourgeois epoch.

The theoretical analyses of Greenberger *et al.* (1989, 1990) and Mermin (1990, 1993) provide a striking example of this phenomenon; see Maudlin (1994) for a detailed analysis of the implications for concepts of causality and temporality. An experimental test, extending the work of Aspect *et al.* (1982), will likely be forthcoming within the next few years.

[26]Bohm (1980). The intimate relations between quantum mechanics and the mind-body problem are discussed in Goldstein (1983, chaps. 7 and 8).

gument is undeniable.[27] In Bohr's words, "Planck's discovery of the *elementary quantum of action* . . . revealed a feature of *wholeness* inherent in atomic physics, going far beyond the ancient idea of the limited divisibility of matter."[28]

Hermeneutics of Classical General Relativity

In the Newtonian mechanistic worldview, space and time are distinct and absolute.[29] In Einstein's special theory of relativity (1905), the distinction between space and time dissolves: there is only a new unity, four-dimensional space-time, and the observer's perception of "space" and "time" depends on her state of motion.[30] In Hermann Minkowski's famous words (1908):

---

[27]Among the voluminous literature, the book by Capra (1975) can be recommended for its scientific accuracy and its accessibility to non-specialists. In addition, the book by Sheldrake (1981), while occasionally speculative, is in general sound. For a sympathetic but critical analysis of New Age theories, see Ross (1991, chap. 1). For a critique of Capra's work from a Third World perspective, see Alvares (1992, chap. 6).

[28]Bohr (1963, p. 2), emphasis in Bohr's original.

[29]Newtonian atomism treats particles as hyperseparated in space and time, backgrounding their interconnectedness (Plumwood 1993a, p. 125); indeed, "the only 'force' allowed within the mechanistic framework is that of kinetic energy—the energy of motion by contact—all other purported forces, including action at a distance, being regarded as occult" (Mathews 1991, p. 17). For critical analyses of the Newtonian mechanistic worldview, see Weil (1968, especially chap. 1), Merchant (1980), Berman (1981), Keller (1985, chaps. 2 and 3), Mathews (1991, chap. 1), and Plumwood (1993a, chap. 5).

[30]According to the traditional textbook account, special relativity is concerned with the coordinate transformations relating *two* frames of reference in uniform relative motion. But this is a misleading oversimplification, as Latour (1988) has pointed out:

> How can one decide whether an observation made in a train about the behaviour of a falling stone can be made to coincide with the observation made of the same falling stone from the embankment? If there are only one, or even *two*, frames of reference, no solution can be found since the man in the train claims he observes a straight line and the man on the embankment a parabola. . . . Einstein's solution is to consider *three* actors: one in the train, one on the embankment and a third one, the author [enunciator] or one of its representants, who tries to superimpose the coded observations sent back by the two others. . . . [W]ithout the enunciator's position (hidden in Einstein's account), and without the notion of centres of calculation, Einstein's own technical argument is ununderstandable . . . [pp. 10–11 and 35, emphasis in original]

In the end, as Latour wittily but accurately observes, special relativity boils down to the proposition that

Henceforth space by itself, and time by itself, are doomed to fade away into mere shadows, and only a kind of union of the two will preserve an independent reality.[31]

Nevertheless, the underlying geometry of Minkowskian space-time remains absolute.[32]

It is in Einstein's general theory of relativity (1915) that the radical conceptual break occurs: the space-time geometry becomes contingent and dynamical, encoding in itself the gravitational field. Mathematically, Einstein breaks with the tradition dating back to Euclid (and which is inflicted on high-school students even today!), and employs instead the non-Euclidean geometry developed by Riemann. Einstein's equations are highly nonlinear, which is why traditionally-trained mathematicians find them so difficult to solve.[33] Newton's gravitational theory corresponds to the crude (and conceptually misleading) truncation of Einstein's equations in which the nonlinearity is simply ignored. Einstein's general relativity therefore subsumes all the putative successes of Newton's theory, while going beyond Newton to predict radically new phenomena that arise directly from the nonlinearity: the bending of starlight by the sun, the precession of the perihelion of Mercury, and the gravitational collapse of stars into black holes.

---

more frames of reference with less privilege can be accessed, reduced, accumulated and combined, observers can be delegated to a few more places in the infinitely large (the cosmos) and the infinitely small (electrons), and the readings they send will be understandable. His [Einstein's] book could well be titled: 'New Instructions for Bringing Back Long-Distance Scientific Travellers'. [pp. 22–23]

Latour's critical analysis of Einstein's logic provides an eminently accessible introduction to special relativity for non-scientists.

[31]Minkowski (1908), translated in Lorentz et al. (1952, p. 75).

[32]It goes without saying that special relativity proposes new concepts not only of space and time but also of mechanics. In special relativity, as Virilio (1991, p. 136) has noted, "the dromospheric space, space-speed, is physically described by what is called the 'logistic equation,' the result of the product of the mass displaced by the speed of its displacement, MxV." This radical alteration of the Newtonian formula has profound consequences, particularly in the quantum theory; see Lorentz et al. (1952) and Weinberg (1992) for further discussion.

[33]Steven Best (1991, p. 225) has put his finger on the crux of the difficulty, which is that "unlike the linear equations used in Newtonian and even quantum mechanics, non-linear equations do [not] have the simple additive property whereby chains of

General relativity is so weird that some of its conse-quences—deduced by impeccable mathematics, and increas-ingly confirmed by astrophysical observation—read like science fiction. Black holes are by now well known, and worm-holes are beginning to make the charts. Perhaps less familiar is Gödel's construction of an Einstein space-time admitting closed timelike curves: that is, a universe in which it is possible to travel *into one's own past!*[34]

Thus, general relativity forces upon us radically new and counterintuitive notions of space, time, and causality[35,36,37,38]; so

---

solutions can be constructed out of simple, independent parts". For this reason, the strategies of atomization, reductionism, and context-stripping that underlie the Newtonian scientific methodology simply do not work in general relativity.

[34]Gödel (1949). For a summary of recent work in this area, see 't Hooft (1993).

[35]These new notions of space, time and causality are *in part* foreshadowed already in special relativity. Thus, Alexander Argyros (1991, p. 137) has noted that

> in a universe dominated by photons, gravitons, and neutrinos, that is, in the very early universe, the theory of special relativity suggests that any distinction between before and after is impossible. For a particle traveling at the speed of light, or one traversing a distance that is in the order of the Planck length, all events are simultaneous.

However, I cannot agree with Argyros' conclusion that Derridean deconstruction is therefore inapplicable to the hermeneutics of early-universe cosmology: Argyros' argument to this effect is based on an impermissibly totalizing use of special relativity (in technical terms, "light-cone coordinates") in a context where *general* relativity is inescapable. (For a similar but less innocent error, see Note 40 below.)

[36]Jean-François Lyotard (1989, pp. 5–6) has pointed out that not only general relativity, but also modern elementary-particle physics, imposes new notions of time:

> In contemporary physics and astrophysics . . . a particle has a sort of elementary memory and consequently a temporal filter. This is why contemporary physicists tend to think that time emanates from matter itself, and that it is not an entity outside or inside the universe whose function it would be to gather all different times into universal history. It is only in certain regions that such—only partial—syntheses could be detected. There would on this view be areas of determinism where complexity is increasing.

Furthermore, Michel Serres (1992, pp. 89–91) has noted that chaos theory (Gleick 1987) and percolation theory (Stauffer 1985) have contested the traditional linear concept of time:

> Time does not always flow along a line . . . or a plane, but along an extraordinarily complex manifold, as if it showed stopping points, ruptures, sinks [*puits*], funnels of overwhelming acceleration [*cheminées d'accélération foudroyante*], rips, lacunae, all sown randomly . . . Time flows in a turbulent and chaotic manner; it percolates. [Translation mine. Note that in the theory of dynamical systems, "*puits*" is a technical term meaning "sink", i.e. the opposite of "source".]

it is not surprising that it has had a profound impact not only on the natural sciences but also on philosophy, literary criticism, and the human sciences. For example, in a celebrated symposium three decades ago on *Les Langages Critiques et les Sciences de l'Homme*, Jean Hyppolite raised an incisive question about Jacques Derrida's theory of structure and sign in scientific discourse:

> When I take, for example, the structure of certain algebraic constructions [ensembles], where is the center? Is the center the knowledge of general rules which, after a fashion, allow us to understand the interplay of the elements? Or is the center certain elements which enjoy a particular privilege within the ensemble? . . . With Einstein, for example, we see the end of a kind of privilege of empiric evidence. And in that connection we see a constant appear, a constant which is a combination of space-time, which does not belong to any of the experimenters who live the experience, but which, in a way, dominates the whole construct; and this notion of the constant— is this the center?[39]

Derrida's perceptive reply went to the heart of classical general relativity:

> The Einsteinian constant is not a constant, is not a center. It is the very concept of variability—it is, finally, the concept of the game. In other words, it is not the concept of some*thing*—

These multiple insights into the nature of time, provided by different branches of physics, are a further illustration of the complementarity principle.

[37]General relativity can arguably be read as corroborating the Nietzschean deconstruction of causality (see e.g. Culler 1982, pp. 86–88), although some relativists find this interpretation problematic. In quantum mechanics, by contrast, this phenomenon is rather firmly established (see Note 25 above).

[38]General relativity is also, of course, the starting point for contemporary astrophysics and physical cosmology. See Mathews (1991, pp. 59–90, 109–116, 142–163) for a detailed analysis of the connections between general relativity (and its generalizations called "geometrodynamics") and an ecological worldview. For an astrophysicist's speculations along similar lines, see Primack and Abrams (1995).

[39]Discussion to Derrida (1970, pp. 265–266).

of a center starting from which an observer could master the
field—but the very concept of the game . . .[40]

In mathematical terms, Derrida's observation relates to the invariance of the Einstein field equation $G_{\mu\nu} = 8\pi G T_{\mu\nu}$ under nonlinear space-time diffeomorphisms (self-mappings of the space-
time manifold which are infinitely differentiable but not
necessarily analytic). The key point is that this invariance group
"acts transitively": this means that any space-time point, if it exists at all, can be transformed into any other. In this way the
infinite-dimensional invariance group erodes the distinction between observer and observed; the π of Euclid and the G of Newton, formerly thought to be constant and universal, are now perceived in their ineluctable historicity; and the putative observer
becomes fatally de-centered, disconnected from any epistemic
link to a space-time point that can no longer be defined by
geometry alone.

Quantum Gravity: String, Weave, or Morphogenetic Field?

However, this interpretation, while adequate within classical
general relativity, becomes incomplete within the emerging
postmodern view of quantum gravity. When even the gravitational field—geometry incarnate—becomes a non-commuting
(and hence nonlinear) operator, how can the classical interpretation of $G_{\mu\nu}$ as a geometric entity be sustained? Now not only

---

[40]Derrida (1970, p. 267). Right-wing critics Gross and Levitt (1994, p. 79) have
ridiculed this statement, willfully misinterpreting it as an assertion about *special*
relativity, in which the Einsteinian constant c (the speed of light in vacuum) is of
course constant. No reader conversant with modern physics—except an
ideologically biased one—could fail to understand Derrida's unequivocal reference to
general relativity.

the observer, but the very concept of geometry, becomes relational and contextual.

The synthesis of quantum theory and general relativity is thus the central unsolved problem of theoretical physics[41]; no one today can predict with confidence what will be the language and ontology, much less the content, of this synthesis, when and if it comes. It is, nevertheless, useful to examine historically the metaphors and imagery that theoretical physicists have employed in their attempts to understand quantum gravity.

The earliest attempts—dating back to the early 1960s—to visualize geometry on the Planck scale (about 10^{-33} centimeters) portrayed it as "space-time foam": bubbles of space-time curvature, sharing a complex and ever-changing topology of interconnections.[42] But physicists were unable to carry this approach farther, perhaps due to the inadequate development at that time of topology and manifold theory (see below).

In the 1970s physicists tried an even more conventional approach: simplify the Einstein equations by pretending that they are *almost linear*, and then apply the standard methods of quantum field theory to the thus-oversimplified equations. But this method, too, failed: it turned out that Einstein's general relativity is, in technical language, "perturbatively nonrenormalizable".[43] This means that the strong nonlinearities of Einstein's

---

[41]Luce Irigaray (1987, pp. 77–78) has pointed out that the contradictions between quantum theory and field theory are in fact the culmination of a historical process that began with Newtonian mechanics:

> The Newtonian break has ushered scientific enterprise into a world where sense perception is worth little, a world which can lead to the annihilation of the very stakes of physics' object: the matter (whatever the predicates) of the universe and of the bodies that constitute it. In this very science, moreover [*d'ailleurs*], cleavages exist: quantum theory/field theory, mechanics of solids/dynamics of fluids, for example. But the imperceptibility of the matter under study often brings with it the paradoxical privilege of *solidity* in discoveries and a delay, even an abandoning of the analysis of the infinity [*l'in-fini*] of the fields of force.

I have here corrected the translation of "*d'ailleurs*", which means "moreover" or "besides" (not "however").

[42]Wheeler (1964).

[43]Isham (1991, sec. 3.1.4).

general relativity are intrinsic to the theory; any attempt to pretend that the nonlinearities are weak is simply self-contradictory. (This is not surprising: the almost-linear approach destroys the most characteristic features of general relativity, such as black holes.)

In the 1980s a very different approach, known as string theory, became popular: here the fundamental constituents of matter are not point-like particles but rather tiny (Planck-scale) closed and open strings.[44] In this theory, the space-time manifold does not exist as an objective physical reality; rather, space-time is a derived concept, an approximation valid only on large length scales (where "large" means "much larger than 10^{-33} centimeters"!). For a while many enthusiasts of string theory thought they were closing in on a Theory of Everything—modesty is not one of their virtues—and some still think so. But the mathematical difficulties in string theory are formidable, and it is far from clear that they will be resolved any time soon.

More recently, a small group of physicists has returned to the full nonlinearities of Einstein's general relativity, and—using a new mathematical symbolism invented by Abhay Ashtekar—they have attempted to visualize the structure of the corresponding quantum theory.[45] The picture they obtain is intriguing: As in string theory, the space-time manifold is only an approximation valid at large distances, not an objective reality. At small (Planck-scale) distances, the geometry of space-time is a *weave*: a complex interconnection of threads.

Finally, an exciting proposal has been taking shape over the past few years in the hands of an interdisciplinary collaboration of mathematicians, astrophysicists, and biologists: this is the theory of the morphogenetic field.[46] Since the mid-1980s ev-

---

[44]Green, Schwarz and Witten (1987).

[45]Ashtekar, Rovelli and Smolin (1992), Smolin (1992).

[46]Sheldrake (1981, 1991), Briggs and Peat (1984, chap. 4), Granero-Porati and Porati (1984), Kazarinoff (1985), Schiffmann (1989), Psarev (1990), Brooks and Castor (1990), Heinonen, Kilpeläinen and Martio (1992), Rensing (1993). For an in-depth treatment of the mathematical background to this theory, see Thom (1975, 1990); and

idence has been accumulating that this field, first conceptualized by developmental biologists[47], is in fact closely linked to the quantum *gravitational* field[48]: (a) it pervades all space; (b) it interacts with all matter and energy, irrespective of whether or not that matter/energy is magnetically charged; and, most significantly, (c) it is what is known mathematically as a "symmetric second-rank tensor". All three properties are characteristic of gravity; and it was proven some years ago that the only self-consistent *nonlinear* theory of a symmetric second-rank tensor field is, at least at low energies, precisely Einstein's general relativity.[49] Thus, if the evidence for (a), (b), and (c) holds up, we can infer that the morphogenetic field is the quantum counterpart of Einstein's gravitational field. Until recently this theory has been ignored or even scorned by the high-energy-physics establishment, who have traditionally resented the encroachment of biologists (not to mention humanists) on their "turf".[50] However, some theoretical physicists have recently begun to give this theory a second look, and there are good prospects for progress in the near future.[51]

---

for a brief but insightful analysis of the philosophical underpinnings of this and related approaches, see Ross (1991, pp. 40–42, 253n).

[47]Waddington (1965), Corner (1966), Gierer *et al.* (1978).

[48]Some early workers thought that the morphogenetic field might be related to the electromagnetic field, but it is now understood that this is merely a suggestive analogy: see Sheldrake (1981, pp. 77, 90) for a clear exposition. Note also point (b) below.

[49]Boulware and Deser (1975).

[50]For another example of the "turf" effect, see Chomsky (1979, pp. 6–7).

[51]To be fair to the high-energy-physics establishment, I should mention that there is also an honest intellectual reason for their opposition to this theory: inasmuch as it posits a subquantum interaction linking patterns throughout the universe, it is, in physicists' terminology, a "non-local field theory". Now, the history of classical theoretical physics since the early 1800s, from Maxwell's electrodynamics to Einstein's general relativity, can be read in a very deep sense as a trend away from action-at-a-distance theories and towards *local field theories:* in technical terms, theories expressible by partial differential equations (Einstein and Infeld 1961, Hayles 1984). So a non-local field theory definitely goes against the grain. On the other hand, as Bell (1987) and others have convincingly argued, the key property of quantum mechanics is precisely its *non-locality*, as expressed in Bell's theorem and its generalizations (see Notes 23 and 24 above). Therefore, a non-local field theory,

It is still too soon to say whether string theory, the space-time weave or morphogenetic fields will be confirmed in the laboratory: the experiments are not easy to perform. But it is intriguing that all three theories have similar conceptual characteristics: strong nonlinearity, subjective space-time, inexorable flux, and a stress on the topology of interconnectedness.

Differential Topology and Homology

Unbeknownst to most outsiders, theoretical physics underwent a significant transformation—albeit not yet a true Kuhnian paradigm shift—in the 1970s and 80s: the traditional tools of mathematical physics (real and complex analysis), which deal with the space-time manifold only locally, were supplemented by topological approaches (more precisely, methods from differential topology[52]) that account for the global (holistic) structure of the universe. This trend was seen in the analysis of anomalies in gauge theories[53]; in the theory of vortex-mediated phase transitions[54]; and in string and superstring theories.[55] Numerous books and review articles on "topology for physicists" were published during these years.[56]

---

although jarring to physicists' classical intuition, is not only natural but in fact *preferred* (and possibly even *mandatory*?) in the quantum context. This is why classical general relativity is a local field theory, while quantum gravity (whether string, weave, or morphogenetic field) is inherently non-local.

[52]Differential topology is the branch of mathematics concerned with those properties of surfaces (and higher-dimensional manifolds) that are unaffected by smooth deformations. The properties it studies are therefore primarily qualitative rather than quantitative, and its methods are holistic rather than Cartesian.

[53]Alvarez-Gaumé (1985). The alert reader will notice that anomalies in "normal science" are the usual harbinger of a *future* paradigm shift (Kuhn 1970).

[54]Kosterlitz and Thouless (1973). The flowering of the theory of phase transitions in the 1970s probably reflects an increased emphasis on discontinuity and rupture in the wider culture: see Note 81 below.

[55]Green, Schwarz and Witten (1987).

[56]A typical such book is Nash and Sen (1983).

At about the same time, in the social and psychological sciences Jacques Lacan pointed out the key role played by differential topology:

This diagram [the Möbius strip] can be considered the basis of a sort of essential inscription at the origin, in the knot which constitutes the subject. This goes much further than you may think at first, because you can search for the sort of surface able to receive such inscriptions. You can perhaps see that the sphere, that old symbol for totality, is unsuitable. A torus, a Klein bottle, a cross-cut surface, are able to receive such a cut. And this diversity is very important as it explains many things about the structure of mental disease. If one can symbolize the subject by this fundamental cut, in the same way one can show that a cut on a torus corresponds to the neurotic subject, and on a cross-cut surface to another sort of mental disease.[57,58]

As Althusser rightly commented, "Lacan finally gives Freud's thinking the scientific concepts that it requires".[59] More recently,

---

[57]Lacan (1970, pp. 192–193), lecture given in 1966. For an in-depth analysis of Lacan's use of ideas from mathematical topology, see Juranville (1984, chap. VII), Granon-Lafont (1985, 1990), Vappereau (1985) and Nasio (1987, 1992); a brief summary is given by Leupin (1991). See Hayles (1990, p. 80) for an intriguing connection between Lacanian topology and chaos theory; unfortunately she does not pursue it. See also Žižek (1991, pp. 38–39, 45–47) for some further homologies between Lacanian theory and contemporary physics. Lacan also made extensive use of concepts from set-theoretic number theory: see e.g. Miller (1977/78) and Ragland-Sullivan (1990).

[58]In bourgeois social psychology, topological ideas had been employed by Kurt Lewin as early as the 1930s, but this work foundered for two reasons: first, because of its individualist ideological preconceptions; and second, because it relied on old-fashioned point-set topology rather than modern differential topology and catastrophe theory. Regarding the second point, see Back (1992).

[59]Althusser (1993, p. 50): "Il suffit, à cette fin, de reconnaître que Lacan confère enfin à la pensée de Freud, les concepts scientifiques qu'elle exige". This famous essay on "Freud and Lacan" was first published in 1964, before Lacan's work had reached its highest level of mathematical rigor. It was reprinted in English translation in *New Left Review* (Althusser 1969).

Lacan's *topologie du sujet* has been applied fruitfully to cinema criticism[60] and to the psychoanalysis of AIDS.[61] In mathematical terms, Lacan is here pointing out that the first homology group[62] of the sphere is trivial, while those of the other surfaces are profound; and this homology is linked with the connectedness or disconnectedness of the surface after one or more cuts.[63] Furthermore, as Lacan suspected, there is an intimate connection between the external structure of the physical world and its inner psychological representation *qua* knot theory: this hypothesis has recently been confirmed by Witten's derivation of knot invariants (in particular the Jones polynomial[64]) from three-dimensional Chern-Simons quantum field theory.[65]

Analogous topological structures arise in quantum gravity, but inasmuch as the manifolds involved are multidimensional rather than two-dimensional, higher homology groups play a role as well. These multidimensional manifolds are no longer amenable to visualization in conventional three-dimensional Cartesian space: for example, the projective space RP^3, which arises from the ordinary 3-sphere by identification of antipodes, would require a Euclidean embedding space of dimension at least 5.[66] Nevertheless, the higher homology groups can be per-

---

[60]Miller (1977/78, especially pp. 24–25). This article has become quite influential in film theory: see e.g. Jameson (1982, pp. 27–28) and the references cited there. As Strathausen (1994, p. 69) indicates, Miller's article is tough going for the reader not well versed in the mathematics of set theory. But it is well worth the effort. For a gentle introduction to set theory, see Bourbaki (1970).

[61]Dean (1993, especially pp. 107–108).

[62]Homology theory is one of the two main branches of the mathematical field called *algebraic topology*. For an excellent introduction to homology theory, see Munkres (1984); or for a more popular account, see Eilenberg and Steenrod (1952). A fully relativistic homology theory is discussed e.g. in Eilenberg and Moore (1965). For a dialectical approach to homology theory and its dual, cohomology theory, see Massey (1978). For a cybernetic approach to homology, see Saludes i Closa (1984).

[63]For the relation of homology to cuts, see Hirsch (1976, pp. 205–208); and for an application to collective movements in quantum field theory, see Caracciolo *et al.* (1993, especially app. A.1).

[64]Jones (1985).

[65]Witten (1989).

[66]James (1971, pp. 271–272). It is, however, worth noting that the space RP^3 is homeomorphic to the group $SO(3)$ of rotational symmetries of conventional three-

ceived, at least approximately, via a suitable multidimensional (nonlinear) logic.[67,68]

Manifold Theory: (W)holes and Boundaries

Luce Irigaray, in her famous article "Is the Subject of Science Sexed?", pointed out that

> the mathematical sciences, in the theory of wholes [*théorie des ensembles*], concern themselves with closed and open spaces . . . They concern themselves very little with the question of the partially open, with wholes that are not clearly delineated [*ensembles flous*], with any analysis of the problem of borders [*bords*] . . .[69]

In 1982, when Irigaray's essay first appeared, this was an incisive criticism: differential topology has traditionally privileged

---

dimensional Euclidean space. Thus, some aspects of three-dimensional Euclidicity are preserved (albeit in modified form) in the postmodern physics, just as some aspects of Newtonian mechanics were preserved in modified form in Einsteinian physics.

[67]Kosko (1993). See also Johnson (1977, pp. 481–482) for an analysis of Derrida's and Lacan's efforts toward transcending the Euclidean spatial logic.

[68]Along related lines, Eve Seguin (1994, p. 61) has noted that "logic says nothing about the world and attributes to the world properties that are but constructs of theoretical thought. This explains why physics since Einstein has relied on alternative logics, such as trivalent logic which rejects the principle of the excluded middle." A pioneering (and unjustly forgotten) work in this direction, likewise inspired by quantum mechanics, is Lupasco (1951). See also Plumwood (1993b, pp. 453–459) for a specifically feminist perspective on nonclassical logics. For a critical analysis of one nonclassical logic ("boundary logic") and its relation to the ideology of cyberspace, see Markley (1994).

[69]Irigaray (1987, pp. 76–77), essay originally appeared in French in 1982. Irigaray's phrase *"théorie des ensembles"* can also be rendered as "theory of sets", and *"bords"* is usually translated in the mathematical context as "boundaries". Her phrase *"ensembles flous"* may refer to the new mathematical field of "fuzzy sets" (Kaufmann 1973, Kosko 1993).

the study of what are known technically as "manifolds without boundary". However, in the past decade, under the impetus of the feminist critique, some mathematicians have given renewed attention to the theory of "manifolds with boundary" [Fr. *variétés à bord*].[70] Perhaps not coincidentally, it is precisely these manifolds that arise in the new physics of conformal field theory, superstring theory, and quantum gravity.

In string theory, the quantum-mechanical amplitude for the interaction of n closed or open strings is represented by a functional integral (basically, a sum) over fields living on a two-dimensional manifold with boundary.[71] In quantum gravity, we may expect that a similar representation will hold, except that the two-dimensional manifold with boundary will be replaced by a multidimensional one. Unfortunately, multidimensionality goes against the grain of conventional linear mathematical thought, and despite a recent broadening of attitudes (notably associated with the study of multidimensional nonlinear phenomena in chaos theory), the theory of multidimensional manifolds with boundary remains somewhat underdeveloped. Nevertheless, physicists' work on the functional-integral approach to quantum gravity continues apace[72], and this work is likely to stimulate the attention of mathematicians.[73]

As Irigaray anticipated, an important question in all of these theories is: Can the boundary be transgressed (crossed), and if so, what happens then? Technically this is known as the problem of "boundary conditions". At a purely mathematical level, the most salient aspect of boundary conditions is the great di-

---

[70]See e.g. Hamza (1990), McAvity and Osborn (1991), Alexander, Berg, and Bishop (1993) and the references cited therein.

[71]Green, Schwarz, and Witten (1987).

[72]Hamber (1992), Nabutosky and Ben-Av (1993), Kontsevich (1994).

[73]In the history of mathematics there has been a long-standing dialectic between the development of its "pure" and "applied" branches (Struik 1987). Of course, the "applications" traditionally privileged in this context have been those profitable to capitalists or useful to their military forces: for example, number theory has been developed largely for its applications in cryptography (Loxton 1990). See also Hardy (1967, pp. 120–121, 131–132).

versity of possibilities: for example, "free b.c." (no obstacle to crossing), "reflecting b.c." (specular reflection as in a mirror), "periodic b.c." (re-entrance in another part of the manifold), and "antiperiodic b.c." (re-entrance with 180° twist). The question posed by physicists is: Of all these conceivable boundary conditions, which ones actually occur in the representation of quantum gravity? Or perhaps, do *all* of them occur simultaneously and on an equal footing, as suggested by the complementarity principle?[74]

At this point my summary of developments in physics must stop, for the simple reason that the answers to these questions— if indeed they have univocal answers—are not yet known. In the remainder of this essay, I propose to take as my starting point those features of the theory of quantum gravity which *are* relatively well established (at least by the standards of conventional science), and attempt to draw out their philosophical and political implications.

Transgressing the Boundaries: Towards a Liberatory Science

Over the past two decades there has been extensive discussion among critical theorists with regard to the characteristics of modernist versus postmodernist culture; and in recent years these dialogues have begun to devote detailed attention to the specific problems posed by the natural sciences.[75] In particular, Madsen and Madsen have recently given a very clear summary

---

[74]The equal representation of all boundary conditions is also suggested by Chew's bootstrap theory of "subatomic democracy": see Chew (1977) for an introduction, and see Morris (1988) and Markley (1992) for philosophical analysis.

[75]Among the large body of works from a diversity of politically progressive perspectives, the books by Merchant (1980), Keller (1985), Harding (1986), Aronowitz (1988b), Haraway (1991), and Ross (1991) have been especially influential. See also the references cited below.

of the characteristics of modernist versus postmodernist science. They posit two criteria for a postmodern science:

> A simple criterion for science to qualify as postmodern is that it be free from any dependence on the concept of objective truth. By this criterion, for example, the complementarity interpretation of quantum physics due to Niels Bohr and the Copenhagen school is seen as postmodernist.[76]

Clearly, quantum gravity is in this respect an archetypal postmodernist science. Secondly,

> The other concept which can be taken as being fundamental to postmodern science is that of *essentiality*. Postmodern scientific theories are constructed from those theoretical elements which are essential for the consistency and utility of the theory.[77]

Thus, quantities or objects which are in principle unobservable—such as space-time points, exact particle positions, or quarks and gluons—ought not to be introduced into the the-

---

[76]Madsen and Madsen (1990, p. 471). The main limitation of the Madsen-Madsen analysis is that it is essentially apolitical; and it hardly needs to be pointed out that disputes over what is *true* can have a profound effect on, and are in turn profoundly affected by, disputes over *political projects*. Thus, Markley (1992, p. 270) makes a point similar to that of Madsen-Madsen, but rightly situates it in its political context:

> Radical critiques of science that seek to escape the constraints of deterministic dialectics must also give over narrowly conceived debates about realism and truth to investigate what kind of realities—political realities—might be engendered by a dialogical bootstrapping. Within a dialogically agitated environment, debates about reality become, in practical terms, irrelevant. "Reality," finally, is a historical construct.

See Markley (1992, pp. 266–272) and Hobsbawm (1993, pp. 63–64) for further discussion of the political implications.

[77]Madsen and Madsen (1990, pp. 471–472).

ory.[78] While much of modern physics is excluded by this criterion, quantum gravity again qualifies: in the passage from classical general relativity to the quantized theory, space-time points
(and indeed the space-time manifold itself) have disappeared
from the theory.

However, these criteria, admirable as they are, are insufficient for a *liberatory* postmodern science: they liberate human
beings from the tyranny of "absolute truth" and "objective reality", but not necessarily from the tyranny of other human beings.

---

[78]Aronowitz (1988b, pp. 292–293) makes a slightly different, but equally cogent,
criticism of quantum chromodynamics (the currently hegemonic theory representing
nucleons as permanently bound states of quarks and gluons): drawing on the work of
Pickering (1984), he notes that

> in his [Pickering's] account, quarks are the name assigned to (absent)
> phenomena that cohere with particle rather than field theories, which, in each
> case, offer different, although equally plausible, explanations for the same
> (inferred) observation. That the majority of the scientific community chose one
> over another is a function of scientists' preference for the tradition rather than
> the validity of explanation.
>
> However, Pickering does not reach back far enough into the history of
> physics to find the basis of the research tradition from which the quark
> explanation emanates. It may not be found inside the tradition but in the
> ideology of science, in the differences behind field versus particle theories,
> simple versus complex explanations, the bias toward certainty rather than
> indeterminateness.

Along very similar lines, Markley (1992, p. 269) observes that physicists' preference
for quantum chromodynamics over Chew's bootstrap theory of "subatomic
democracy" (Chew 1977) is a result of ideology rather than data:

> It is not surprising, in this regard, that bootstrap theory has fallen into relative
> disfavor among physicists seeking a GUT (Grand Unified Theory) or TOE
> (Theory of Everything) to explain the structure of the universe. Comprehensive
> theories that explain "everything" are products of the privileging of coherence
> and order in western science. The choice between bootstrap theory and theories
> of everything that confronts physicists does *not* have to do primarily with the
> truth-value offered by these accounts of available data but with the narrative
> structures—indeterminate or deterministic—into which these data are placed
> and by which they are interpreted.

Unfortunately, the vast majority of physicists are not yet aware of these incisive
critiques of one of their most fervently-held dogmas.

For another critique of the hidden ideology of contemporary particle physics,
see Kroker *et al.* (1989, pp. 158–162, 204–207). The style of this critique is rather too
Baudrillardian for my staid taste, but the content is (except for a few minor
inaccuracies) right on target.

In Andrew Ross' words, we need a science "that will be publicly answerable and of some service to progressive interests."[79] From a feminist standpoint, Kelly Oliver makes a similar argument:

> . . . in order to be revolutionary, feminist theory cannot claim to describe what exists, or, "natural facts." Rather, feminist theories should be political tools, strategies for overcoming oppression in specific concrete situations. The goal, then, of feminist theory, should be to develop *strategic* theories—not true theories, not false theories, but strategic theories.[80]

How, then, is this to be done?

In what follows, I would like to discuss the outlines of a liberatory postmodern science on two levels: first, with regard to general themes and attitudes; and second, with regard to political goals and strategies.

One characteristic of the emerging postmodern science is its stress on nonlinearity and discontinuity: this is evident, for example, in chaos theory and the theory of phase transitions as well as in quantum gravity.[81] At the same time, feminist thinkers have pointed out the need for an adequate analysis of fluidity, in particular turbulent fluidity.[82] These two themes are not as contradictory as it might at first appear: turbulence connects with

---

[79]Ross (1991, p. 29). For an amusing example of how this modest demand has driven right-wing scientists into fits of apoplexy ("frighteningly Stalinist" is the chosen epithet), see Gross and Levitt (1994, p. 91).

[80]Oliver (1989, p. 146).

[81]While chaos theory has been deeply studied by cultural analysts—see e.g. Hayles (1990, 1991), Argyros (1991), Best (1991), Young (1991, 1992), Assad (1993), among many others—the theory of phase transitions has passed largely unremarked. (One exception is the discussion of the renormalization group in Hayles (1990, pp. 154–158).) This is a pity, because discontinuity and the emergence of multiple scales are central features in this theory; and it would be interesting to know how the development of these themes in the 1970s and afterwards is connected to trends in the wider culture. I therefore suggest this theory as a fruitful field for future research by cultural analysts. Some theorems on discontinuity which may be relevant to this analysis can be found in Van Enter, Fernández, and Sokal (1993).

[82]Irigaray (1985), Hayles (1992). See, however, Schor (1989) for a critique of Irigaray's undue deference toward conventional (male) science, particularly physics.

strong nonlinearity, and smoothness/fluidity is sometimes associated with discontinuity (e.g. in catastrophe theory[83]); so a synthesis is by no means out of the question.

Secondly, the postmodern sciences deconstruct and transcend the Cartesian metaphysical distinctions between humankind and Nature, observer and observed, Subject and Object. Already quantum mechanics, earlier in this century, shattered the ingenuous Newtonian faith in an objective, prelinguistic world of material objects "out there"; no longer could we ask, as Heisenberg put it, whether "particles exist in space and time objectively". But Heisenberg's formulation still presupposes the objective existence of space and time as the neutral, unproblematic arena in which quantized particle-waves interact (albeit indeterministically); and it is precisely this would-be arena that quantum gravity problematizes. Just as quantum mechanics informs us that the position and momentum of a particle are brought into being only by the act of observation, so quantum gravity informs us that space and time themselves are contextual, their meaning defined only relative to the mode of observation.[84]

Thirdly, the postmodern sciences overthrow the static ontological categories and hierarchies characteristic of modernist science. In place of atomism and reductionism, the new sciences stress the dynamic web of relationships between the whole and the part; in place of fixed individual essences (e.g.

---

[83] Thom (1975, 1990), Arnol'd (1992).

[84] Concerning the Cartesian/Baconian metaphysics, Robert Markley (1991, p. 6) has observed that

> Narratives of scientific progress depend upon imposing binary oppositions—true/false, right/wrong—on theoretical and experimental knowledge, privileging meaning over noise, metonymy over metaphor, monological authority over dialogical contention. . . . [T]hese attempts to fix nature are ideologically coercive as well as descriptively limited. They focus attention only on the small range of phenomena—say, linear dynamics—which seem to offer easy, often idealized ways of modeling and interpreting humankind's relationship to the universe.

While this observation is informed primarily by chaos theory—and secondarily by nonrelativistic quantum mechanics—it in fact summarizes beautifully the radical challenge to modernist metaphysics posed by quantum gravity.

Newtonian particles), they conceptualize interactions and flows (e.g. quantum fields). Intriguingly, these homologous features arise in numerous seemingly disparate areas of science, from quantum gravity to chaos theory to the biophysics of self-organizing systems. In this way, the postmodern sciences appear to be converging on a new epistemological paradigm, one that may be termed an *ecological* perspective, broadly understood as "recogniz[ing] the fundamental interdependence of all phenomena and the embeddedness of individuals and societies in the cyclical patterns of nature."[85]

A fourth aspect of postmodern science is its self-conscious stress on symbolism and representation. As Robert Markley points out, the postmodern sciences are increasingly transgressing disciplinary boundaries, taking on characteristics that had heretofore been the province of the humanities:

> Quantum physics, hadron bootstrap theory, complex number theory, and chaos theory share the basic assumption that reality cannot be described in linear terms, that nonlinear—and unsolvable—equations are the only means possible to describe a complex, chaotic, and non-deterministic reality. These postmodern theories are—significantly—all meta-critical in the sense that they foreground themselves as metaphors rather than as "accurate" descriptions of reality. In terms that are more familiar to literary theorists than to theoretical physicists, we might say that these attempts by scientists to develop new strategies of description represent notes towards a theory of theories, of how representation—mathematical, experimental, and verbal—is inherently com-

[85]Capra (1988, p. 145). One caveat: I have strong reservations about Capra's use here of the word "cyclical", which if interpreted too literally could promote a politically regressive quietism. For further analyses of these issues, see Bohm (1980), Merchant (1980, 1992), Berman (1981), Prigogine and Stengers (1984), Bowen (1985), Griffin (1988), Kitchener (1988), Callicott (1989, chaps. 6 and 9), Shiva (1990), Best (1991), Haraway (1991, 1994), Mathews (1991), Morin (1992), Santos (1992), and Wright (1992).

plex and problematizing, not a solution but part of the semiotics of investigating the universe.[86,87]

From a different starting point, Aronowitz likewise suggests that a liberatory science may arise from interdisciplinary sharing of epistemologies:

> . . . natural objects are also socially constructed. It is not a question of whether these natural objects, or, to be more precise, the objects of natural scientific knowledge, exist independently of the act of knowing. This question is answered by the assumption of "real" time as opposed to the presupposition, common among neo-Kantians, that time always has a referent, that temporality is therefore a relative, not an unconditioned, category. Surely, the earth evolved long before life on earth. The question is whether objects of natural scientific knowledge are constituted outside the social field. If this is possible, we can assume that science or art may develop procedures that effectively neutralize the effects emanating from the means by which we produce knowledge/art. Performance art may be such an attempt.[88]

Finally, postmodern science provides a powerful refutation of the authoritarianism and elitism inherent in traditional science, as well as an empirical basis for a democratic approach to scientific work. For, as Bohr noted, "a complete elucidation of one and the same object may require diverse points of view which defy a unique description"—this is quite simply a fact

---

[86]Markley (1992, p. 264). A minor quibble: It is not clear to me that complex number theory, which is a new and still quite speculative branch of mathematical physics, ought to be accorded the same epistemological status as the three firmly established sciences cited by Markley.

[87]See Wallerstein (1993, pp. 17–20) for an incisive and closely analogous account of how the postmodern physics is beginning to borrow ideas from the historical social sciences; and see Santos (1989, 1992) for a more detailed development.

[88]Aronowitz (1988b, p. 344).

about the world, much as the self-proclaimed empiricists of modernist science might prefer to deny it. In such a situation, how can a self-perpetuating secular priesthood of credentialed "scientists" purport to maintain a monopoly on the production of scientific knowledge? (Let me emphasize that I am in no way opposed to specialized scientific training; I object only when an elite caste seeks to impose its canon of "high science", with the aim of excluding *a priori* alternative forms of scientific production by nonmembers.[89])

The content and methodology of postmodern science thus provide powerful intellectual support for the progressive political project, understood in its broadest sense: the transgressing of boundaries, the breaking down of barriers, the radical democratization of all aspects of social, economic, political, and cultural life.[90] Conversely, one part of this project must involve the construction of a new and truly progressive science that can serve the needs of such a democratized society-to-be. As Markley observes, there seem to be two more-or-less mutually exclusive choices available to the progressive community:

---

[89]At this point, the traditional scientist's response is that work not conforming to the evidentiary standards of conventional science is fundamentally *irrational*, i.e. logically flawed and therefore not worthy of credence. But this refutation is insufficient: for, as Porush (1993) has lucidly observed, modern mathematics and physics have *themselves* admitted a powerful "intrusion of the irrational" in quantum mechanics and Gödel's theorem—although, understandably, like the Pythagoreans 24 centuries ago, modernist scientists have attempted to exorcise this unwanted irrational element as best they could. Porush makes a powerful plea for a "post-rational epistemology" that would retain the best of conventional Western science while validating alternative ways of knowing.

Note also that Jacques Lacan, from a quite different starting point, came long ago to a similar appreciation of the inevitable role of irrationality in modern mathematics:

> If you'll permit me to use one of those formulas which come to me as I write my notes, human life could be defined as a calculus in which zero was irrational. This formula is just an image, a mathematical metaphor. When I say "irrational," I'm referring not to some unfathomable emotional state but precisely to what is called an imaginary number. The square root of minus one doesn't correspond to anything that is subject to our intuition, anything real—in the mathematical sense of the term—and yet, it must be conserved, along with its full function.

[Lacan (1977, pp. 28–29), seminar originally given in 1959.]

For further reflections on irrationality in modern mathematics, see Solomon (1988, p. 76) and Bloor (1991, pp. 122–125).

[90]See e.g. Aronowitz (1994) and the discussion following it.

On the one hand, politically progressive scientists can try to recuperate existing practices for moral values they uphold, arguing that their right-wing enemies are defacing nature and that they, the counter-movement, have access to the truth. [But] the state of the biosphere—air pollution, water pollution, disappearing rain forests, thousands of species on the verge of extinction, large areas of land burdened far beyond their carrying capacity, nuclear power plants, nuclear weapons, clearcuts where there used to be forests, starvation, malnutrition, disappearing wetlands, nonexistent grass lands, and a rash of environmentally caused diseases—suggests that the realist dream of scientific progress, of recapturing rather than revolutionizing existing methodologies and technologies, is, at worst, irrelevant to a political struggle that seeks something more than a reenactment of state socialism.[91]

The alternative is a profound reconception of science as well as politics:

[T]he dialogical move towards redefining systems, of seeing the world not only as an ecological whole but as a set of competing systems—a world held together by the tensions among various natural and human interests—offers the possibility of redefining what science is and what it does, of restructuring deterministic schemes of scientific education in favor of ongoing dialogues about how we intervene in our environment.[92]

It goes without saying that postmodernist science unequivocally favors the latter, deeper approach.

---

[91]Markley (1992, p. 271).

[92]Markley (1992, p. 271). Along parallel lines, Donna Haraway (1991, pp. 191–192) has argued eloquently for a democratic science comprising "partial, locatable, critical knowledges sustaining the possibility of webs of connections called solidarity in politics and shared conversations in epistemology" and founded on "a doctrine and practice of objectivity that privileges contestation, deconstruction, passionate construction, webbed connections, and hope for transformation of systems of knowledge and ways of seeing." These ideas are further developed in Haraway (1994) and Doyle (1994).

In addition to redefining the content of science, it is imperative to restructure and redefine the institutional loci in which scientific labor takes place—universities, government labs, and corporations—and reframe the reward system that pushes scientists to become, often against their own better instincts, the hired guns of capitalists and the military. As Aronowitz has noted, "One third of the 11,000 physics graduate students in the United States are in the single subfield of solid state physics, and all of them will be able to get jobs in that subfield."[93] By contrast, there are few jobs available in either quantum gravity or environmental physics.

But all this is only a first step: the fundamental goal of any emancipatory movement must be to demystify and democratize the production of scientific knowledge, to break down the artificial barriers that separate "scientists" from "the public". Realistically, this task must start with the younger generation, through a profound reform of the educational system.[94] The teaching of science and mathematics must be purged of its authoritarian and elitist characteristics[95], and the content of these subjects enriched by incorporating the insights of the feminist[96], queer[97], multiculturalist[98], and ecological[99] critiques.

---

[93]Aronowitz (1988b, p. 351). Although this observation appeared in 1988, it is all the more true today.

[94]Freire (1970), Aronowitz, and Giroux (1991, 1993).

[95]For an example in the context of the Sandinista revolution, see Sokal (1987).

[96]Merchant (1980), Easlea (1981), Keller (1985, 1992), Harding (1986, 1991), Haraway (1989, 1991), Plumwood (1993a). See Wylie *et al.* (1990) for an extensive bibliography. The feminist critique of science has, not surprisingly, been the object of a bitter right-wing counterattack. For a sampling, see Levin (1988), Haack (1992, 1993), Sommers (1994), Gross and Levitt (1994, chap. 5), and Patai and Koertge (1994).

[97]Trebilcot (1988), Hamill (1994).

[98]Ezeabasili (1977), Van Sertima (1983), Frye (1987), Sardar (1988), Adams (1990), Nandy (1990), Alvares (1992), Harding (1994). As with the feminist critique, the multiculturalist perspective has been ridiculed by right-wing critics, with a condescension that in some cases borders on racism. See e.g. Ortiz de Montellano (1991), Martel (1991/92), Hughes (1993, chap. 2) and Gross and Levitt (1994, pp. 203–214).

[99]Merchant (1980, 1992), Berman (1981), Callicott (1989, chaps. 6 and 9), Mathews (1991), Wright (1992), Plumwood (1993a), Ross (1994).

Finally, the content of any science is profoundly constrained by the language within which its discourses are formulated; and mainstream Western physical science has, since Galileo, been formulated in the language of mathematics.[100,101] But *whose*

---

[100]See Wojciehowski (1991) for a deconstruction of Galileo's rhetoric, in particular his claim that the mathematico-scientific method can lead to direct and reliable knowledge of "reality".

[101]A very recent but important contribution to the philosophy of mathematics can be found in the work of Deleuze and Guattari (1994, chap. 5). Here they introduce the philosophically fruitful notion of a "functive" [Fr. *fonctif*], which is neither a function [Fr. *fonction*] nor a functional [Fr. *fonctionnelle*] but rather a more basic conceptual entity:

> The object of science is not concepts but rather functions that are presented as propositions in discursive systems. The elements of functions are called *functives*. [p. 117]

This apparently simple idea has surprisingly subtle and far-reaching consequences; its elucidation requires a detour into chaos theory (see also Rosenberg 1993 and Canning 1994):

> . . . the first difference between science and philosophy is their respective attitudes toward chaos. Chaos is defined not so much by its disorder as by the infinite speed with which every form taking shape in its vanishes. It is a void that is not a nothingness but a *virtual*, containing all possible particles and drawing out all possible forms, which spring up only to disappear immediately, without consistency or reference, without consequence. Chaos is an infinite speed of birth and disappearance. [pp. 117–118]

But science, unlike philosophy, cannot cope with infinite speeds:

> . . . it is by slowing down that matter, as well as the scientific thought able to penetrate it [*sic*] with propositions, is actualized. A function is a Slow-motion. Of course, science constantly advances accelerations, not only in catalysis but in particle accelerators and expansions that move galaxies apart. However, the primordial slowing down is not for these phenomena a zero-instant with which they break but rather a condition coextensive with their whole development. To slow down is to set a limit in chaos to which all speeds are subject, so that they form a variable determined as abscissa, at the same time as the limit forms a universal constant that cannot be gone beyond (for example, a maximum degree of contraction). *The first functives are therefore the limit and the variable*, and reference is a relationship between values of the variable or, more profoundly, the relationship of the variable, as abscissa of speeds, with the limit. [pp. 118–119, emphasis mine]

A rather intricate further analysis (too lengthy to quote here) leads to a conclusion of profound methodological importance for those sciences based on mathematical modelling:

> The respective independence of variables appears in mathematics when one of them is at a higher power than the first. That is why Hegel shows that variability in the function is not confined to values that can be changed (2/3 and 4/6) or are left undetermined ($a = 2b$) but requires one of the variables to be at a higher power ($y^2/x = P$). [p. 122]

(Note that the English translation inadvertently writes $y^{2/x} = P$, an amusing error that thoroughly mangles the logic of the argument.)

mathematics? The question is a fundamental one, for, as
Aronowitz has observed, "neither logic nor mathematics escapes the 'contamination' of the social."[102] And as feminist
thinkers have repeatedly pointed out, in the present culture this
contamination is overwhelmingly capitalist, patriarchal, and
militaristic: "mathematics is portrayed as a woman whose nature desires to be the conquered Other."[103,104] Thus, a liberatory

---

Surprisingly for a technical philosophical work, this book *(Qu'est-ce que la
philosophie?)* was a best-seller in France in 1991. It has recently appeared in English
translation, but is, alas, unlikely to compete successfully with Rush Limbaugh and
Howard Stern for the best-seller lists in this country.

[102]Aronowitz (1988b, p. 346). For a vicious right-wing attack on this proposition, see
Gross and Levitt (1994, pp. 52–54). See Ginzberg (1989), Cope-Kasten (1989), Nye
(1990), and Plumwood (1993b) for lucid feminist critiques of conventional
(masculinist) mathematical logic, in particular the *modus ponens* and the syllogism.
Concerning the *modus ponens*, see also Woolgar (1988, pp. 45–46) and Bloor (1991,
p. 182); and concerning the syllogism, see also Woolgar (1988, pp. 47–48) and Bloor
(1991, pp. 131–135). For an analysis of the social images underlying mathematical
conceptions of infinity, see Harding (1986, p. 50). For a demonstration of the social
contextuality of mathematical statements, see Woolgar (1988, p. 43) and Bloor (1991,
pp. 107–130).

[103]Campbell and Campbell-Wright (1995, p. 135). See Merchant (1980) for a detailed
analysis of the themes of control and domination in Western mathematics and
science.

[104]Let me mention in passing two other examples of sexism and militarism in
mathematics that to my knowledge have not been noticed previously:
    The first concerns the theory of branching processes, which arose in Victorian
England from the "problem of the extinction of families", and which now plays a key
role *inter alia* in the analysis of nuclear chain reactions (Harris 1963). In the seminal
(and this sexist word is apt) paper on the subject, Francis Galton and the Reverend
H.W. Watson wrote (1874):

> The decay of the families of men who occupied conspicuous positions in
> past times has been a subject of frequent research, and has given rise to various
> conjectures . . . The instances are very numerous in which surnames that were
> once common have since become scarce or have wholly disappeared. The
> tendency is universal, and, in explanation of it, the conclusion has hastily been
> drawn that a rise in physical comfort and intellectual capacity is necessarily
> accompanied by a diminution in 'fertility' . . .
> Let $p_0, p_1, p_2, \ldots$ be the respective probabilities that a man has 0, 1, 2, . . .
> sons, let each son have the same probability of sons of his own, and so on. What
> is the probability that the male line is extinct after r generations, and more
> generally what is the probability for any given number of descendants in the
> male line in any given generation?

One cannot fail to be charmed by the quaint implication that human males reproduce
asexually; nevertheless, the classism, social-Darwinism, and sexism in this passage
are obvious.
    The second example is Laurent Schwartz's 1973 book on *Radon Measures*.
While technically quite interesting, this work is imbued, as its title makes plain, with
the pro-nuclear-energy worldview that has been characteristic of French science

science cannot be complete without a profound revision of the canon of mathematics.[105] As yet no such emancipatory mathematics exist, and we can only speculate upon its eventual content. We can see hints of it in the multidimensional and nonlinear logic of fuzzy systems theory[106]; but this approach is still heavily marked by its origins in the crisis of late-capitalist production relations.[107] Catastrophe theory[108], with its dialectical emphases on smoothness/discontinuity and metamorphosis/unfolding, will indubitably play a major role in the future mathematics; but much theoretical work remains to be done before this approach can become a concrete tool of progressive political praxis.[109] Finally, chaos theory—which provides our deepest insights into the ubiquitous yet mysterious phenomenon of nonlinearity—will be central to all future mathematics. And yet, these images of the future mathematics must remain but the haziest glimmer: for, alongside these three young branches in the tree of science, there will arise new trunks and branches— entire new theoretical frameworks—of which we, with our present ideological blinders, cannot yet even conceive.

I wish to thank Giacomo Caracciolo, Lucía Fernández-Santoro, Lia Gutiérrez, and Elizabeth Meiklejohn for enjoyable discussions which have contributed greatly to this article.

---

since the early 1960s. Sadly, the French left—especially but by no means solely the PCF—has traditionally been as enthusiastic for nuclear energy as the right (see Touraine *et al.* 1980).

[105]Just as liberal feminists are frequently content with a minimal agenda of legal and social equality for women and "pro-choice", so liberal (and even some socialist) mathematicians are often content to work within the hegemonic Zermelo-Fraenkel framework (which, reflecting its nineteenth-century liberal origins, already incorporates the axiom of equality) supplemented only by the axiom of choice. But this framework is grossly insufficient for a liberatory mathematics, as was proven long ago by Cohen (1966).

[106]Kosko (1993).

[107]Fuzzy systems theory has been heavily developed by transnational corporations— first in Japan and later elsewhere—to solve practical problems of efficiency in labor-displacing automation.

[108]Thom (1975, 1990), Arnol'd (1992).

[109]An interesting start is made by Schubert (1989).

Needless to say, these people should not be assumed to be in total agreement with the scientific and political views expressed here, nor are they responsible for any errors or obscurities which may inadvertently remain.

Works Cited

Adams, Hunter Havelin III. 1990. African and African-American contributions to science and technology. In *African-American Baseline Essays*. Portland, Ore.: Multnomah School District 1J, Portland Public Schools.

Albert, David Z. 1992. *Quantum Mechanics and Experience*. Cambridge: Harvard University Press.

Alexander, Stephanie B., I. David Berg and Richard L. Bishop. 1993. Geometric curvature bounds in Riemannian manifolds with boundary. *Transactions of the American Mathematical Society* **339**: 703–716.

Althusser, Louis. 1969. Freud and Lacan. *New Left Review* **55**: 48–65.

Althusser, Louis. 1993. *Écrits sur la psychanalyse: Freud et Lacan*. Paris: Stock/IMEC.

Alvares, Claude. 1992. *Science, Development and Violence: The Revolt against Modernity*. Delhi: Oxford University Press.

Alvarez-Gaumé, Luís. 1985. Topology and anomalies. In *Mathematics and Physics: Lectures on Recent Results*, vol. 2, pp. 50–83, edited by L. Streit. Singapore: World Scientific.

Argyros, Alexander J. 1991. *A Blessed Rage for Order: Deconstruction, Evolution, and Chaos*. Ann Arbor: University of Michigan Press.

Arnol'd, Vladimir I. 1992. *Catastrophe Theory*. 3rd ed. Translated by G.S. Wassermann and R.K. Thomas. Berlin: Springer.

Aronowitz, Stanley. 1981. *The Crisis in Historical Materialism: Class, Politics and Culture in Marxist Theory*. New York: Praeger.

Aronowitz, Stanley. 1988a. The production of scientific knowledge: Science, ideology, and Marxism. In *Marxism and the Interpretation of Culture*, pp. 519–541, edited by Cary Nelson and Lawrence Grossberg. Urbana and Chicago: University of Illinois Press.

Aronowitz, Stanley. 1988b. *Science as Power: Discourse and Ideology in Modern Society*. Minneapolis: University of Minnesota Press.

Aronowitz, Stanley. 1994. The situation of the left in the United States. *Socialist Review* **23**(3): 5–79.

Aronowitz, Stanley and Henry A. Giroux. 1991. *Postmodern Education: Politics, Culture, and Social Criticism*. Minneapolis: University of Minnesota Press.

Aronowitz, Stanley and Henry A. Giroux. 1993. *Education Still Under Siege*. Westport, Conn.: Bergin & Garvey.

Ashtekar, Abhay, Carlo Rovelli and Lee Smolin. 1992. Weaving a classical metric with quantum threads. *Physical Review Letters* **69**: 237–240.

Aspect, Alain, Jean Dalibard and Gérard Roger. 1982. Experimental test of Bell's inequalities using time-varying analyzers. *Physical Review Letters* **49**: 1804–1807.

Assad, Maria L. 1993. Portrait of a nonlinear dynamical system: The discourse of Michel Serres. *SubStance* **71/72**: 141–152.

Back, Kurt W. 1992. This business of topology. *Journal of Social Issues* **48**(2): 51–66.

Bell, John S. 1987. *Speakable and Unspeakable in Quantum Mechanics: Collected Papers on Quantum Philosophy*. New York: Cambridge University Press.

Berman, Morris. 1981. *The Reenchantment of the World*. Ithaca, N.Y.: Cornell University Press.

Best, Steven. 1991. Chaos and entropy: Metaphors in postmodern science and social theory. *Science as Culture* **2**(2) (no. 11): 188–226.

Bloor, David. 1991. *Knowledge and Social Imagery*. 2nd ed. Chicago: University of Chicago Press.

Bohm, David. 1980. *Wholeness and the Implicate Order*. London: Routledge & Kegan Paul.

Bohr, Niels. 1958. Natural philosophy and human cultures. In *Essays 1932–1957 on Atomic Physics and Human Knowledge* (The Philosophical Writings of Niels Bohr, Volume II), pp. 23–31. New York: Wiley.

Bohr, Niels. 1963. Quantum physics and philosophy—causality and complementarity. In *Essays 1958–1962 on Atomic Physics and Human Knowledge* (The Philosophical Writings of Niels Bohr, Volume III), pp. 1–7. New York: Wiley.

Booker, M. Keith. 1990. Joyce, Planck, Einstein, and Heisenberg: A relativistic quantum mechanical discussion of *Ulysses*. *James Joyce Quarterly* **27**: 577–586.

Boulware, David G. and S. Deser. 1975. Classical general relativity derived from quantum gravity. *Annals of Physics* **89**: 193–240.

Bourbaki, Nicolas. 1970. *Théorie des ensembles*. Paris: Hermann.

Bowen, Margarita. 1985. The ecology of knowledge: Linking the natural and social sciences. *Geoforum* **16**: 213–225.

Bricmont, Jean. 1994. Contre la philosophie de la mécanique quantique. Texte d'une communication faite au colloque "Faut-il promouvoir les échanges entre les sciences et la philosophie?", Louvain-la-Neuve (Belgium), 24–25 mars 1994. [Published in R. Franck, ed., *Les Sciences et la philosophie. Quatorze essais de rapprochement*, pp. 131–179, Paris, Vrin, 1995.]

Briggs, John and F. David Peat. 1984. *Looking Glass Universe: The Emerging Science of Wholeness*. New York: Cornerstone Library.

Brooks, Roger and David Castor. 1990. Morphisms between supersymmetric and topological quantum field theories. *Physics Letters B* **246**: 99–104.

Callicott, J. Baird. 1989. *In Defense of the Land Ethic: Essays in Environmental Philosophy*. Albany, N.Y.: State University of New York Press.

Campbell, Mary Anne and Randall K. Campbell-Wright. 1995. Toward a feminist algebra. In *Teaching the Majority: Science, Mathematics, and Engineering That Attracts Women*, edited by Sue V. Rosser. New York: Teachers College Press.

Canning, Peter. 1994. The crack of time and the ideal game. In *Gilles Deleuze and the Theater of Philosophy*, pp. 73–98, edited by Constantin V. Boundas and Dorothea Olkowski. New York: Routledge.

Capra, Fritjof. 1975. *The Tao of Physics: An Exploration of the Parallels Between Modern Physics and Eastern Mysticism*. Berkeley, Calif.: Shambhala.

Capra, Fritjof. 1988. The role of physics in the current change of paradigms. In *The World View of Contemporary Physics: Does It Need a New Metaphysics?*, pp. 144–155, edited by Richard F. Kitchener. Albany, N.Y.: State University of New York Press.

Caracciolo, Sergio, Robert G. Edwards, Andrea Pelissetto and Alan D. Sokal. 1993. Wolff-type embedding algorithms for general nonlinear σ-models. *Nuclear Physics B* **403**: 475–541.

Chew, Geoffrey. 1977. Impasse for the elementary-particle concept. In *The Sciences Today*, pp. 366–399, edited by Robert M. Hutchins and Mortimer Adler. New York: Arno Press.

Chomsky, Noam. 1979. *Language and Responsibility*. Translated by John Viertel. New York: Pantheon.

Cohen, Paul J. 1966. *Set Theory and the Continuum Hypothesis*. New York: Benjamin.

Coleman, Sidney. 1993. Quantum mechanics in your face. Lecture at New York University, November 12, 1993.

Cope-Kasten, Vance. 1989. A portrait of dominating rationality. *Newsletters on Computer Use, Feminism, Law, Medicine, Teaching (American Philosophical Association)* **88**(2) (March): 29–34.

Corner, M.A. 1966. Morphogenetic field properties of the forebrain area of the neural plate in an anuran. *Experientia* **22**: 188–189.

Craige, Betty Jean. 1982. *Literary Relativity: An Essay on Twentieth-Century Narrative*. Lewisburg: Bucknell University Press.

Culler, Jonathan. 1982. *On Deconstruction: Theory and Criticism after Structuralism*. Ithaca, N.Y.: Cornell University Press.

Dean, Tim. 1993. The psychoanalysis of AIDS. *October* **63**: 83–116.

Deleuze, Gilles and Félix Guattari. 1994. *What is Philosophy?* Translated by Hugh Tomlinson and Graham Burchell. New York: Columbia University Press.

Derrida, Jacques. 1970. Structure, sign and play in the discourse of the human sciences. In *The Languages of Criticism and the Sciences of Man: The Structuralist Con-

troversy, pp. 247–272, edited by Richard Macksey and Eugenio Donato. Baltimore: Johns Hopkins Press.

Doyle, Richard. 1994. Dislocating knowledge, thinking out of joint: Rhizomatics, *Caenorhabditis elegans* and the importance of being multiple. *Configurations: A Journal of Literature, Science, and Technology* **2**: 47–58.

Dürr, Detlef, Sheldon Goldstein and Nino Zanghí. 1992. Quantum equilibrium and the origin of absolute uncertainty. *Journal of Statistical Physics* **67**: 843–907.

Easlea, Brian. 1981. *Science and Sexual Oppression: Patriarchy's Confrontation with Women and Nature.* London: Weidenfeld and Nicolson.

Eilenberg, Samuel and John C. Moore. 1965. *Foundations of Relative Homological Algebra.* Providence, R.I.: American Mathematical Society.

Eilenberg, Samuel and Norman E. Steenrod. 1952. *Foundations of Algebraic Topology.* Princeton, N.J.: Princeton University Press.

Einstein, Albert and Leopold Infeld. 1961. *The Evolution of Physics.* New York: Simon and Schuster.

Ezeabasili, Nwankwo. 1977. *African Science: Myth or Reality?* New York: Vantage Press.

Feyerabend, Paul K. 1975. *Against Method: Outline of an Anarchistic Theory of Knowledge.* London: New Left Books.

Freire, Paulo. 1970. *Pedagogy of the Oppressed.* Translated by Myra Bergman Ramos. New York: Continuum.

Froula, Christine. 1985. Quantum physics/postmodern metaphysics: The nature of Jacques Derrida. *Western Humanities Review* **39**: 287–313.

Frye, Charles A. 1987. Einstein and African religion and philosophy: The hermetic parallel. In *Einstein and the Humanities*, pp. 59–70, edited by Dennis P. Ryan. New York: Greenwood Press.

Galton, Francis and H.W. Watson. 1874. On the probability of the extinction of families. *Journal of the Anthropological Institute of Great Britain and Ireland* **4**: 138–144.

Gierer, A., R.C. Leif, T. Maden and J.D. Watson. 1978. Physical aspects of generation of morphogenetic fields and tissue forms. In *Differentiation and Development*, edited by F. Ahmad, J. Schultz, T.R. Russell and R. Werner. New York: Academic Press.

Ginzberg, Ruth. 1989. Feminism, rationality, and logic. *Newsletters on Computer Use, Feminism, Law, Medicine, Teaching (American Philosophical Association)* **88**(2) (March): 34–39.

Gleick, James. 1987. *Chaos: Making a New Science.* New York: Viking.

Gödel, Kurt. 1949. An example of a new type of cosmological solutions of Einstein's field equations of gravitation. *Reviews of Modern Physics* **21**: 447–450.

Goldstein, Rebecca. 1983. *The Mind-Body Problem.* New York: Random House.

Granero-Porati, M.I. and A. Porati. 1984. Temporal organization in a morphogenetic field. *Journal of Mathematical Biology* **20**: 153–157.

Granon-Lafont, Jeanne. 1985. *La Topologie ordinaire de Jacques Lacan*. Paris: Point Hors Ligne.

Granon-Lafont, Jeanne. 1990. *Topologie lacanienne et clinique analytique*. Paris: Point Hors Ligne.

Green, Michael B., John H. Schwarz and Edward Witten. 1987. *Superstring Theory*. 2 vols. New York: Cambridge University Press.

Greenberg, Valerie D. 1990. *Transgressive Readings: The Texts of Franz Kafka and Max Planck*. Ann Arbor: University of Michigan Press.

Greenberger, D.M., M.A. Horne and Z. Zeilinger. 1989. Going beyond Bell's theorem. In *Bell's Theorem, Quantum Theory and Conceptions of the Universe*, pp. 73–76, edited by M. Kafatos. Dordrecht: Kluwer.

Greenberger, D.M., M.A. Horne, A. Shimony and Z. Zeilinger. 1990. Bell's theorem without inequalities. *American Journal of Physics* **58**: 1131–1143.

Griffin, David Ray, ed. 1988. *The Reenchantment of Science: Postmodern Proposals*. Albany, N.Y.: State University of New York Press.

Gross, Paul R. and Norman Levitt. 1994. *Higher Superstition: The Academic Left and its Quarrels with Science*. Baltimore: Johns Hopkins University Press.

Haack, Susan. 1992. Science 'from a feminist perspective'. *Philosophy* **67**: 5–18.

Haack, Susan. 1993. Epistemological reflections of an old feminist. *Reason Papers* **18** (fall): 31–43.

Hamber, Herbert W. 1992. Phases of four-dimensional simplicial quantum gravity. *Physical Review D* **45**: 507–512.

Hamill, Graham. 1994. The epistemology of expurgation: Bacon and *The Masculine Birth of Time*. In *Queering the Renaissance*, pp. 236–252, edited by Jonathan Goldberg. Durham, N.C.: Duke University Press.

Hamza, Hichem. 1990. Sur les transformations conformes des variétés riemanniennes à bord. *Journal of Functional Analysis* **92**: 403–447.

Haraway, Donna J. 1989. *Primate Visions: Gender, Race, and Nature in the World of Modern Science*. New York: Routledge.

Haraway, Donna J. 1991. *Simians, Cyborgs, and Women: The Reinvention of Nature*. New York: Routledge.

Haraway, Donna J. 1994. A game of cat's cradle: Science studies, feminist theory, cultural studies. *Configurations: A Journal of Literature, Science, and Technology* **2**: 59–71.

Harding, Sandra. 1986. *The Science Question in Feminism*. Ithaca, N.Y.: Cornell University Press.

Harding, Sandra. 1991. *Whose Science? Whose Knowledge? Thinking from Women's Lives*. Ithaca, N.Y.: Cornell University Press.

Harding, Sandra. 1994. Is science multicultural? Challenges, resources, opportunities, uncertainties. *Configurations: A Journal of Literature, Science, and Technology* **2**: 301–330.

Hardy, G.H. 1967. *A Mathematician's Apology.* Cambridge: Cambridge University Press.

Harris, Theodore E. 1963. *The Theory of Branching Processes.* Berlin: Springer.

Hayles, N. Katherine. 1984. *The Cosmic Web: Scientific Field Models and Literary Strategies in the Twentieth Century.* Ithaca, N.Y.: Cornell University Press.

Hayles, N. Katherine. 1990. *Chaos Bound: Orderly Disorder in Contemporary Literature and Science.* Ithaca, N.Y.: Cornell University Press.

Hayles, N. Katherine, ed. 1991. *Chaos and Order: Complex Dynamics in Literature and Science.* Chicago: University of Chicago Press.

Hayles, N. Katherine. 1992. Gender encoding in fluid mechanics: Masculine channels and feminine flows. *Differences: A Journal of Feminist Cultural Studies* **4**(2): 16–44.

Heinonen, J., T. Kilpeläinen and O. Martio. 1992. Harmonic morphisms in nonlinear potential theory. *Nagoya Mathematical Journal* **125**: 115–140.

Heisenberg, Werner. 1958. *The Physicist's Conception of Nature.* Translated by Arnold J. Pomerans. New York: Harcourt, Brace.

Hirsch, Morris W. 1976. *Differential Topology.* New York: Springer.

Hobsbawm, Eric. 1993. The new threat to history. *New York Review of Books* (16 December): 62–64.

Hochroth, Lysa. 1995. The scientific imperative: Improductive expenditure and energeticism. *Configurations: A Journal of Literature, Science, and Technology* **3**: 47–77.

Honner, John. 1994. Description and deconstruction: Niels Bohr and modern philosophy. In *Niels Bohr and Contemporary Philosophy* (Boston Studies in the Philosophy of Science #153), pp. 141–153, edited by Jan Faye and Henry J. Folse. Dordrecht: Kluwer.

Hughes, Robert. 1993. *Culture of Complaint: The Fraying of America.* New York: Oxford University Press.

Irigaray, Luce. 1985. The 'mechanics' of fluids. In *This Sex Which Is Not One.* Translated by Catherine Porter with Carolyn Burke. Ithaca, N.Y.: Cornell University Press.

Irigaray, Luce. 1987. Le sujet de la science est-il sexué? / Is the subject of science sexed? Translated by Carol Mastrangelo Bové. *Hypatia* **2**(3): 65–87.

Isham, C.J. 1991. Conceptual and geometrical problems in quantum gravity. In *Recent Aspects of Quantum Fields* (Lecture Notes in Physics #396), edited by H. Mitter and H. Gausterer. Berlin: Springer.

Itzykson, Claude and Jean-Bernard Zuber. 1980. *Quantum Field Theory.* New York: McGraw-Hill International.

James, I.M. 1971. Euclidean models of projective spaces. *Bulletin of the London Mathematical Society* **3**: 257–276.

Jameson, Fredric. 1982. Reading Hitchcock. *October* **23**: 15–42.

Jammer, Max. 1974. *The Philosophy of Quantum Mechanics.* New York: Wiley.

Johnson, Barbara. 1977. The frame of reference: Poe, Lacan, Derrida. *Yale French Studies* **55/56**: 457–505.

Johnson, Barbara. 1989. *A World of Difference.* Baltimore: Johns Hopkins University Press.

Jones, V.F.R. 1985. A polynomial invariant for links via Von Neumann algebras. *Bulletin of the American Mathematical Society* **12**: 103–112.

Juranville, Alain. 1984. *Lacan et la philosophie.* Paris: Presses Universitaires de France.

Kaufmann, Arnold. 1973. *Introduction à la théorie des sous-ensembles flous à l'usage des ingénieurs.* Paris: Masson.

Kazarinoff, N.D. 1985. Pattern formation and morphogenetic fields. In *Mathematical Essays on Growth and the Emergence of Form*, pp. 207–220, edited by Peter L. Antonelli. Edmonton: University of Alberta Press.

Keller, Evelyn Fox. 1985. *Reflections on Gender and Science.* New Haven: Yale University Press.

Keller, Evelyn Fox. 1992. *Secrets of Life, Secrets of Death: Essays on Language, Gender, and Science.* New York: Routledge.

Kitchener, Richard F., ed. 1988. *The World View of Contemporary Physics: Does It Need a New Metaphysics?* Albany, N.Y.: State University of New York Press.

Kontsevich, M. 1994. Résultats rigoureux pour modèles sigma topologiques. Conférence au XIème Congrès International de Physique Mathématique, Paris, 18–23 juillet 1994. Edité par Daniel Iagolnitzer et Jacques Toubon. À paraître.

Kosko, Bart. 1993. *Fuzzy Thinking: The New Science of Fuzzy Logic.* New York: Hyperion.

Kosterlitz, J.M. and D.J. Thouless. 1973. Ordering, metastability and phase transitions in two-dimensional systems. *Journal of Physics C* **6**: 1181–1203.

Kroker, Arthur, Marilouise Kroker and David Cook. 1989. *Panic Encyclopedia: The Definitive Guide to the Postmodern Scene.* New York: St. Martin's Press.

Kuhn, Thomas S. 1970. *The Structure of Scientific Revolutions.* 2nd ed. Chicago: University of Chicago Press.

Lacan, Jacques. 1970. Of structure as an inmixing of an otherness prerequisite to any subject whatever. In *The Languages of Criticism and the Sciences of Man*, pp. 186–200, edited by Richard Macksey and Eugenio Donato. Baltimore: Johns Hopkins Press.

Lacan, Jacques. 1977. Desire and the interpretation of desire in *Hamlet*. Translated by James Hulbert. *Yale French Studies* **55/56**: 11–52.

Latour, Bruno. 1987. *Science in Action: How to Follow Scientists and Engineers Through Society.* Cambridge, Mass.: Harvard University Press.

Latour, Bruno. 1988. A relativistic account of Einstein's relativity. *Social Studies of Science* **18**: 3–44.

Leupin, Alexandre. 1991. Introduction: Voids and knots in knowledge and truth. In *Lacan and the Human Sciences*, pp. 1–23, edited by Alexandre Leupin. Lincoln, Neb.: University of Nebraska Press.

Levin, Margarita. 1988. Caring new world: Feminism and science. *American Scholar* **57**: 100–106.

Lorentz, H.A., A. Einstein, H. Minkowski, and H. Weyl. 1952. *The Principle of Relativity.* Translated by W. Perrett and G.B. Jeffery. New York: Dover.

Loxton, J.H., ed. 1990. *Number Theory and Cryptography.* Cambridge–New York: Cambridge University Press.

Lupasco, Stéphane. 1951. *Le Principe d'antagonisme et la logique de l'énergie.* Actualités Scientifiques et Industrielles #1133. Paris: Hermann.

Lyotard, Jean-François. 1989. Time today. Translated by Geoffrey Bennington and Rachel Bowlby. *Oxford Literary Review* **11**: 3–20.

Madsen, Mark and Deborah Madsen. 1990. Structuring postmodern science. *Science and Culture* **56**: 467–472.

Markley, Robert. 1991. What now? An introduction to interphysics. *New Orleans Review* **18**(1): 5–8.

Markley, Robert. 1992. The irrelevance of reality: Science, ideology and the postmodern universe. *Genre* **25**: 249–276.

Markley, Robert. 1994. Boundaries: Mathematics, alienation, and the metaphysics of cyberspace. *Configurations: A Journal of Literature, Science, and Technology* **2**: 485–507.

Martel, Erich. 1991/92. How valid are the Portland baseline essays? *Educational Leadership* **49**(4): 20–23.

Massey, William S. 1978. *Homology and Cohomology Theory.* New York: Marcel Dekker.

Mathews, Freya. 1991. *The Ecological Self.* London: Routledge.

Maudlin, Tim. 1994. *Quantum Non-Locality and Relativity: Metaphysical Intimations of Modern Physics.* Aristotelian Society Series, vol. 13. Oxford: Blackwell.

McAvity, D.M. and H. Osborn. 1991. A DeWitt expansion of the heat kernel for manifolds with a boundary. *Classical and Quantum Gravity* **8**: 603–638.

McCarthy, Paul. 1992. Postmodern pleasure and perversity: Scientism and sadism. *Postmodern Culture* **2**, no. 3. Available as mccarthy.592 from listserv@listserv.ncsu.edu or http://jefferson.village.virginia.edu/pmc (Internet). Also reprinted in *Essays in Postmodern Culture*, pp. 99–132, edited by Eyal Amiran and John Unsworth. New York: Oxford University Press, 1993.

Merchant, Carolyn. 1980. *The Death of Nature: Women, Ecology, and the Scientific Revolution.* New York: Harper & Row.

Merchant, Carolyn. 1992. *Radical Ecology: The Search for a Livable World.* New York: Routledge.

Mermin, N. David. 1990. Quantum mysteries revisited. *American Journal of Physics* **58**: 731–734.

Mermin, N. David. 1993. Hidden variables and the two theorems of John Bell. *Reviews of Modern Physics* **65**: 803–815.

Merz, Martina and Karin Knorr Cetina. 1994. Deconstruction in a 'thinking' science: Theoretical physicists at work. Geneva: European Laboratory for Particle Physics (CERN), preprint CERN-TH.7152/94. [Published in *Social Studies of Science* **27** (1997): 73–111.]

Miller, Jacques-Alain. 1977/78. Suture (elements of the logic of the signifier). *Screen* **18**(4): 24–34.

Morin, Edgar. 1992. *The Nature of Nature* (Method: Towards a Study of Humankind, vol. 1). Translated by J.L. Roland Bélanger. New York: Peter Lang.

Morris, David B. 1988. Bootstrap theory: Pope, physics, and interpretation. *The Eighteenth Century: Theory and Interpretation* **29**: 101–121.

Munkres, James R. 1984. *Elements of Algebraic Topology.* Menlo Park, Calif.: Addison-Wesley.

Nabutosky, A. and R. Ben-Av. 1993. Noncomputability arising in dynamical triangulation model of four-dimensional quantum gravity. *Communications in Mathematical Physics* **157**: 93–98.

Nandy, Ashis, ed. 1990. *Science, Hegemony and Violence: A Requiem for Modernity.* Delhi: Oxford University Press.

Nash, Charles and Siddhartha Sen. 1983. *Topology and Geometry for Physicists.* London: Academic Press.

Nasio, Juan-David. 1987. *Les Yeux de Laure: Le concept d'objet "a" dans la théorie de J. Lacan. Suivi d'une introduction à la topologie psychanalytique.* Paris: Aubier.

Nasio, Juan-David. 1992. Le concept de sujet de l'inconscient. Texte d'une intervention realisée dans le cadre du séminaire de Jacques Lacan "La topologie et le temps", le mardi 15 mai 1979. In *Cinq leçons sur la théorie de Jacques Lacan.* Paris: Éditions Rivages.

Nye, Andrea. 1990. *Words of Power: A Feminist Reading of the History of Logic.* New York: Routledge.

Oliver, Kelly. 1989. Keller's gender/science system: Is the philosophy of science to science as science is to nature? *Hypatia* **3**(3): 137–148.

Ortiz de Montellano, Bernard. 1991. Multicultural pseudoscience: Spreading scientific illiteracy among minorities: Part I. *Skeptical Inquirer* **16**(2): 46–50.

Overstreet, David. 1980. Oxymoronic language and logic in quantum mechanics and James Joyce. *Sub-Stance* **28**: 37–59.

Pais, Abraham. 1991. *Niels Bohr's Times: In Physics, Philosophy, and Polity.* New York: Oxford University Press.

Patai, Daphne and Noretta Koertge. 1994. *Professing Feminism: Cautionary Tales from the Strange World of Women's Studies.* New York: Basic Books.

Pickering, Andrew. 1984. *Constructing Quarks: A Sociological History of Particle Physics.* Chicago: University of Chicago Press.

Plotnitsky, Arkady. 1994. *Complementarity: Anti-Epistemology after Bohr and Derrida.* Durham, N.C.: Duke University Press.

Plumwood, Val. 1993a. *Feminism and the Mastery of Nature.* London: Routledge.

Plumwood, Val. 1993b. The politics of reason: Towards a feminist logic. *Australasian Journal of Philosophy* 71: 436–462.

Porter, Jeffrey. 1990. "Three quarks for Muster Mark": Quantum wordplay and nuclear discourse in Russell Hoban's *Riddley Walker. Contemporary Literature* 21: 448–469.

Porush, David. 1989. Cybernetic fiction and postmodern science. *New Literary History* 20: 373–396.

Porush, David. 1993. Voyage to Eudoxia: The emergence of a post-rational epistemology in literature and science. *SubStance* 71/72: 38–49.

Prigogine, Ilya and Isabelle Stengers. 1984. *Order out of Chaos: Man's New Dialogue with Nature.* New York: Bantam.

Primack, Joel R. and Nancy Ellen Abrams. 1995. "In a beginning . . .": Quantum cosmology and Kabbalah. *Tikkun* 10(1) (January/February): 66–73.

Psarev, V.I. 1990. Morphogenesis of distributions of microparticles by dimensions in the coarsening of dispersed systems. *Soviet Physics Journal* 33: 1028–1033.

Ragland-Sullivan, Ellie. 1990. Counting from 0 to 6: Lacan, "suture", and the imaginary order. In *Criticism and Lacan: Essays and Dialogue on Language, Structure, and the Unconscious*, pp. 31–63, edited by Patrick Colm Hogan and Lalita Pandit. Athens, Ga.: University of Georgia Press.

Rensing, Ludger, ed. 1993. Oscillatory signals in morphogenetic fields. Part II of *Oscillations and Morphogenesis*, pp. 133–209. New York: Marcel Dekker.

Rosenberg, Martin E. 1993. Dynamic and thermodynamic tropes of the subject in Freud and in Deleuze and Guattari. *Postmodern Culture* 4, no. 1. Available as rosenber.993 from listserv@listserv.ncsu.edu or http://jefferson.village.virginia.edu/pmc (Internet).

Ross, Andrew. 1991. *Strange Weather: Culture, Science, and Technology in the Age of Limits.* London: Verso.

Ross, Andrew. 1994. *The Chicago Gangster Theory of Life: Nature's Debt to Society.* London: Verso.

Saludes i Closa, Jordi. 1984. Un programa per a calcular l'homologia simplicial. *Butlletí de la Societat Catalana de Ciències* (segona època) 3: 127–146.

Santos, Boaventura de Sousa. 1989. *Introdução a uma Ciência Pós-Moderna.* Porto: Edições Afrontamento.

Santos, Boaventura de Sousa. 1992. A discourse on the sciences. *Review (Fernand Braudel Center)* **15**(1): 9–47.

Sardar, Ziauddin, ed. 1988. *The Revenge of Athena: Science, Exploitation and the Third World.* London: Mansell.

Schiffmann, Yoram. 1989. The second messenger system as the morphogenetic field. *Biochemical and Biophysical Research Communications* **165**: 1267–1271.

Schor, Naomi. 1989. This essentialism which is not one: Coming to grips with Irigaray. *Differences: A Journal of Feminist Cultural Studies* **1**(2): 38–58.

Schubert, G. 1989. Catastrophe theory, evolutionary extinction, and revolutionary politics. *Journal of Social and Biological Structures* **12**: 259–279.

Schwartz, Laurent. 1973. *Radon Measures on Arbitrary Topological Spaces and Cylindrical Measures.* London: Oxford University Press.

Seguin, Eve. 1994. A modest reason. *Theory, Culture & Society* **11**(3): 55–75.

Serres, Michel. 1992. *Éclaircissements: Cinq entretiens avec Bruno Latour.* Paris: François Bourin.

Sheldrake, Rupert. 1981. *A New Science of Life: The Hypothesis of Formative Causation.* Los Angeles: J.P. Tarcher.

Sheldrake, Rupert. 1991. *The Rebirth of Nature.* New York: Bantam.

Shiva, Vandana. 1990. Reductionist science as epistemological violence. In *Science, Hegemony and Violence: A Requiem for Modernity*, pp. 232–256, edited by Ashis Nandy. Delhi: Oxford University Press.

Smolin, Lee. 1992. Recent developments in nonperturbative quantum gravity. In *Quantum Gravity and Cosmology* (Proceedings 1991, Sant Feliu de Guixols, Estat Lliure de Catalunya), pp. 3–84, edited by J. Pérez-Mercader, J. Sola and E. Verdaguer. Singapore: World Scientific.

Sokal, Alan D. 1982. An alternate constructive approach to the φ_3^4 quantum field theory, and a possible destructive approach to φ_4^4. *Annales de l'Institut Henri Poincaré A* **37**: 317–398.

Sokal, Alan. 1987. Informe sobre el plan de estudios de las carreras de Matemática, Estadística y Computación. Report to the Universidad Nacional Autónoma de Nicaragua, Managua, unpublished.

Solomon, J. Fisher. 1988. *Discourse and Reference in the Nuclear Age.* Oklahoma Project for Discourse and Theory, vol. 2. Norman: University of Oklahoma Press.

Sommers, Christina Hoff. 1994. *Who Stole Feminism?: How Women Have Betrayed Women.* New York: Simon & Schuster.

Stauffer, Dietrich. 1985. *Introduction to Percolation Theory.* London: Taylor & Francis.

Strathausen, Carsten. 1994. Althusser's mirror. *Studies in 20th Century Literature* **18:** 61–73.

Struik, Dirk Jan. 1987. *A Concise History of Mathematics.* 4th rev. ed. New York: Dover.

Thom, René. 1975. *Structural Stability and Morphogenesis.* Translated by D.H. Fowler. Reading, Mass.: Benjamin.

Thom, René. 1990. *Semio Physics: A Sketch.* Translated by Vendla Meyer. Redwood City, Calif.: Addison-Wesley.

't Hooft, G. 1993. Cosmology in 2+1 dimensions. *Nuclear Physics B (Proceedings Supplement)* **30:** 200–203.

Touraine, Alain, Zsuzsa Hegedus, François Dubet and Michel Wievorka. 1980. *La Prophétie anti-nucléaire.* Paris: Éditions du Seuil.

Trebilcot, Joyce. 1988. Dyke methods, or Principles for the discovery/creation of the withstanding. *Hypatia* **3**(2): 1–13.

Van Enter, Aernout C.D., Roberto Fernández and Alan D. Sokal. 1993. Regularity properties and pathologies of position-space renormalization-group transformations: Scope and limitations of Gibbsian theory. *Journal of Statistical Physics* **72:** 879–1167.

Van Sertima, Ivan, ed. 1983. *Blacks in Science: Ancient and Modern.* New Brunswick, N.J.: Transaction Books.

Vappereau, Jean Michel. 1985. *Essaim: Le Groupe fondamental du noeud.* Psychanalyse et Topologie du Sujet. Paris: Point Hors Ligne.

Virilio, Paul. 1991. *The Lost Dimension.* Translation of *L'espace critique.* Translated by Daniel Moshenberg. New York: Semiotext(e).

Waddington, C.H. 1965. Autogenous cellular periodicities as (a) temporal templates and (b) basis of 'morphogenetic fields'. *Journal of Theoretical Biology* **8:** 367–369.

Wallerstein, Immanuel. 1993. The TimeSpace of world-systems analysis: A philosophical essay. *Historical Geography* **23**(1/2): 5–22.

Weil, Simone. 1968. *On Science, Necessity, and the Love of God.* Translated and edited by Richard Rees. London: Oxford University Press.

Weinberg, Steven. 1992. *Dreams of a Final Theory.* New York: Pantheon.

Wheeler, John A. 1964. Geometrodynamics and the issue of the final state. In *Relativity, Groups and Topology,* edited by Cécile M. DeWitt and Bryce S. DeWitt. New York: Gordon and Breach.

Witten, Edward. 1989. Quantum field theory and the Jones polynomial. *Communications in Mathematical Physics* **121:** 351–399.

Wojciehowski, Dolora Ann. 1991. Galileo's two chief word systems. *Stanford Italian Review* **10:** 61–80.

Woolgar, Steve. 1988. *Science: The Very Idea.* Chichester, England: Ellis Horwood.

Wright, Will. 1992. *Wild Knowledge: Science, Language, and Social Life in a Fragile Environment.* Minneapolis: University of Minnesota Press.

Wylie, Alison, Kathleen Okruhlik, Sandra Morton and Leslie Thielen-Wilson. 1990. Philosophical feminism: A bibliographic guide to critiques of science. *Resources for Feminist Research/Documentation sur la Recherche Féministe* **19**(2) (June): 2–36.

Young, T.R. 1991. Chaos theory and symbolic interaction theory: Poetics for the postmodern sociologist. *Symbolic Interaction* **14:** 321–334.

Young, T.R. 1992. Chaos theory and human agency: Humanist sociology in a postmodern era. *Humanity & Society* **16:** 441–460.

Žižek, Slavoj. 1991. *Looking Awry: An Introduction to Jacques Lacan through Popular Culture.* Cambridge, Mass.: MIT Press.

B. Some Comments on the Parody

Let us note first that all the references cited in the parody are real, and all the quotes are rigorously accurate; nothing has been invented (unfortunately). The text constantly illustrates what David Lodge calls "a law of academic life: *it is impossible to be excessive in flattery of one's peers.*"[110]

The purpose of the following remarks is to explain some of the tricks used in constructing the parody, to indicate what exactly is being spoofed in certain passages, and to clarify our position with respect to those ideas. This last point is particularly important, as it is in the nature of a parody to conceal the author's true views. (Indeed, in many cases Sokal parodied extreme or ambiguously stated versions of ideas that he in fact holds in more nuanced and precisely stated forms.) However, we do not have the space to explain everything, and we shall leave to the reader the pleasure of discovering many other jokes hidden in the text.

Introduction

The article's first two paragraphs set forth an extraordinarily radical version of social constructivism, culminating in the claim that physical reality (and not merely our ideas about it) is "at bottom a social and linguistic construct". The goal in these

---

[110]Lodge (1984, p. 152), italics in the original.

paragraphs was not to summarize the views of the *Social Text* editors—much less those of the authors cited in notes 1–3— but to test whether the bald assertion (without evidence or argument) of such an extreme thesis would raise any eyebrows among the editors. If it did, they never bothered to communicate their misgivings to Sokal, despite his repeated requests for comments, criticisms, and suggestions. See Chapter 4 for our real views on these matters.

The works praised in this section are dubious at best. Quantum mechanics is *not* primarily the product of a "cultural fabric", but the reference to a work by one of *Social Text*'s editors (Aronowitz) couldn't hurt. Ditto for the reference to Ross: here "oppositional discourses in post-quantum science" is a euphemism for channeling, crystal therapy, morphogenetic fields, and sundry other New Age enthusiasms. Irigaray's and Hayles' exegeses of "gender encoding in fluid mechanics" are analyzed in Chapter 5.

To say that space-time ceases to be an objective reality in quantum gravity is premature for two reasons. Firstly, a complete theory of quantum gravity does not yet exist, so we do not know what it will imply. Secondly, though quantum gravity *will* very likely entail radical changes in our concepts of space and time—they may, for example, cease to be fundamental elements in the theory, and become instead an approximate description valid on scales greater than 10^{-33} centimeters[111]—this does not mean that space-time stops being objective, except in the banal sense that tables and chairs are not "objective" because they are composed of atoms. Finally, it is exceedingly unlikely that a theory about space-time on subatomic scales could have valid *political* implications!

Note, in passing, the use of postmodernist jargon: "problematized", "relativized", and so forth (in particular, about existence itself).

---

[111]This is ten trillion trillion (10^{25}) times smaller than an atom.

Quantum Mechanics

This section exemplifies two aspects of postmodernist musings on quantum mechanics: first, a tendency to confuse the technical meanings of words such as "uncertainty" or "discontinuity" with their everyday meanings; and second, a fondness for the most subjectivist writings of Heisenberg and Bohr, interpreted in a radical way that goes far beyond their own views (which are in turn vigorously disputed by many physicists and philosophers of science). But postmodern philosophy loves the multiplicity of viewpoints, the importance of the observer, holism, and indeterminism. For a *serious* discussion of the philosophical problems posed by quantum mechanics, see the references listed in note 8 (in particular, Albert's book is an excellent introduction for non-experts).

Note 13 on Porush is a joke on vulgar economism. In fact, all contemporary technology is based on semiconductor physics, which in turn depends in crucial ways on quantum mechanics.

McCarthy's "thought-provoking analysis" (note 20) begins as follows:

> This study traces the nature and consequences of the circulation of desire in a postmodern order of things (an order implicitly modelled on a repressed archetype of the new physics' fluid particle flows), and it reveals a complicity between scientism, which underpins the postmodern condition, and the sadism of incessant deconstruction, which heightens the intensity of the pleasure-seeking moment in postmodernism.

The rest of the article is in the same vein.

Aronowitz's text (note 25) is a web of confusions and it would take too much space to disentangle them all. Suffice it to say that the problems raised by quantum mechanics (and in particular by Bell's theorem) have little to do with "time's reversal" and nothing at all to do with time's "segmentation into hours and minutes" or "industrial discipline in the early bourgeois epoch".

Goldstein's book on the mind-body problem (note 26) is an enjoyable *novel.*

Capra's speculations on the link between quantum mechanics and Oriental philosophy are, in our view, dubious to say the least. Sheldrake's theory of "morphogenetic fields", though popular in New Age circles, hardly qualifies as "in general sound".

Hermeneutics of Classical General Relativity

The references to physics in this section and the next are, by and large, roughly correct though incredibly shallow; they are written in a deliberately overblown style that parodies some recent popularizations of science. Nevertheless, the text is riddled with absurdities. For example, Einstein's nonlinear equations are indeed difficult to solve, especially for those who do *not* have a "traditional" mathematical training. This reference to "nonlinearity" is the start of a recurrent joke, which imitates the misunderstandings rife in postmodernist writings (see p. 143–45 above). Wormholes and Gödel's space-time are rather speculative theoretical ideas; one of the defects of much contemporary scientific popularization is, in fact, to put the best-established and the most speculative aspects of physics on the same footing.

The notes contain several delights. The quotes from Latour (note 30) and Virilio (note 32) are analyzed in Chapters 6 and 10, respectively. Lyotard's text (note 36) mixes together the terminology of at least three distinct branches of physics—elementary-particle physics, cosmology, and chaos and complexity theory—in a completely arbitrary way. Serres' rhapsody on chaos theory (note 36) confuses the state of the system, which can move in a complex and unpredictable way (see Chapter 7), with the nature of time itself, which flows in the conventional manner ("along a line"). Furthermore, percolation theory deals with the flow of fluids in porous media[112] and says nothing about the nature of space and time.

But the primary purpose of this section is to provide a gentle lead-in to the article's first major gibberish quote, namely Derrida's comment on relativity ("the Einsteinian constant is not a constant . . ."). We haven't the slightest idea what this means—and neither, apparently, does Derrida—but as it is a one-shot abuse, committed orally at a conference, we shall not belabor the point.[113] The paragraph following the Derrida quote, which exhibits a gradual crescendo of absurdity, is one of our favorites. It goes without saying that a mathematical constant such as π does not change over time, even if our ideas about it may.

Quantum Gravity

The first major blooper in this section concerns the expression "noncommuting (and hence nonlinear)". In actual fact, quantum mechanics uses noncommuting operators that are perfectly *linear*. This joke is inspired by a text of Markley quoted later in the article (p. 238).

The next five paragraphs provide a superficial, but essentially correct, overview of physicists' attempts to construct a theory of quantum gravity. Note, however, the exaggerated emphasis on "metaphors and imagery", "nonlinearity", "flux", and "interconnectedness".

The enthusiastic reference to the morphogenetic field is, by contrast, completely arbitrary. Nothing in contemporary science

---

[112]See, for example, de Gennes (1976).

[113]For an amusing attempt, by a postmodernist author who does know some physics, to come up with something Derrida's words could conceivably have meant that might make sense, see Plotnitsky (1997). The trouble is that Plotnitsky comes up with at least *two* alternative technical interpretations of Derrida's phrase "the Einsteinian constant", without providing any convincing evidence that Derrida intended (or even understood) either of them.

can be invoked to support this New Age fantasy which, in any case, has nothing to do with quantum gravity. Sokal was led to this "theory" by the favorable allusion of Ross (note 46), one of the editors of *Social Text*.

The reference to Chomsky on the "turf" effect (note 50) was dangerous, as the editors could very well have known this text or looked it up. It is the one we quote in the Introduction (note 11 on p. 12), and it says essentially the opposite of what is suggested in the parody.

The discussion of non-locality in quantum mechanics (note 51) is deliberately confused, but since this problem is rather technical, we can only refer the reader, for example, to Maudlin's book.

Note, finally, the illogic embodied in the expression "subjective space-time". The fact that space-time may cease to be a fundamental entity in a future theory of quantum gravity does not make it in any way "subjective".

Differential Topology

This section contains the article's second major piece of authoritative nonsense, namely Lacan's text on psychoanalytic topology (which we analyze in Chapter 2). The articles applying Lacanian topology to film criticism and the psychoanalysis of AIDS are, sadly, real. Knot theory does indeed have beautiful applications in contemporary physics—as Witten and others have shown—but this has nothing to do with Lacan.

The last paragraph plays on the postmodern fondness for "multidimensionality" and "nonlinearity" by inventing a nonexistent field: "multidimensional (nonlinear) logic".

Manifold Theory

The quote from Irigaray is discussed in Chapter 5. The parody again suggests that "conventional" science has an aversion to anything that is "multidimensional"; but the truth is that *all* interesting manifolds are multidimensional.[114] Manifolds with boundary are a classic subject of differential geometry.

Note 73 is deliberately exaggerated, though we are sympathetic to the idea that economic and political power struggles strongly affect how science gets translated into technology and for whose benefit. Cryptography does indeed have military (as well as commercial) applications and has in recent years become increasingly based on number theory. However, number theory has fascinated mathematicians since antiquity, and until recently it had very few "practical" applications of any kind: it was the branch of pure mathematics *par excellence*. The reference to Hardy was dangerous: in this very accessible autobiography, he prides himself on working in mathematical fields that have no applications. (There is an additional irony in this reference. Writing in 1941, Hardy listed two branches of science that, in his view, will never have military applications: number theory and Einstein's relativity. Futurology is a risky enterprise, indeed!)

Towards a Liberatory Science

This section combines gross confusions about science with exceedingly sloppy thinking about philosophy and politics. Nevertheless, it also contains some ideas—on the link between scientists and the military, on ideological bias in science, on the pedagogy of science—with which we partly agree, at least when these ideas are formulated more carefully. We do not want the parody to provoke unqualified derision toward these ideas, and we refer the reader to the Epilogue for our real views on some of them.

This section begins by claiming that "postmodern" science has freed itself from objective truth. But, whatever opinions sci-

---

[114]"Manifold" is a geometrical concept that generalizes the notion of surface to spaces of more than two dimensions.

entists may have on chaos or quantum mechanics, they clearly do not consider themselves "liberated" from the goal of objectivity; were that the case, they would simply have ceased to do science. Nevertheless, a whole book would be needed to disentangle the confusions concerning chaos, quantum physics, and self-organization that underlie this sort of idea; see Chapter 7 for a brief analysis.

Having freed science from the goal of objectivity, the article then proposes to politicize science in the worst sense, judging scientific theories not by their correspondence to reality but by their compatibility with one's ideological preconceptions. The quote from Kelly Oliver, which makes this politicization explicit, raises the perennial problem of self-refutation: how can one know whether or not a theory is "strategic", except by asking whether it is *truly*, *objectively* efficacious in promoting one's declared political goals? The problems of truth and objectivity cannot be evaded so easily. Similarly, Markley's claim (" 'Reality', finally is a historical construct", note 76) is both philosophically confused and politically pernicious: it opens the door to the worst nationalist and religious-fundamentalist excesses, as Hobsbawm eloquently demonstrates (p. 207–8).

Here are, finally, some glaring absurdities in this section:

—Markley (p. 238) puts complex number theory—which, in fact, goes back at least to the early nineteenth century and belongs to mathematics, not physics—in the same bag as quantum mechanics, chaos theory, and the now-largely-defunct hadron bootstrap theory. He has probably confused it with the recent, and very speculative, theories on *complexity*. Note 86 is an ironic joke at his expense.

—Many of the 11,000 graduate students working in solid-state physics would be pleasantly surprised to learn that they will all find jobs in their subfield (p. 242).

—The word "Radon" in the title of Schwartz's book (note 104) is the name of a mathematician. The book deals with pure mathematics and has nothing to do with nuclear energy.

—The axiom of equality (note 105) says that two sets are equal if and only if they have the same elements. To link this axiom with nineteenth-century liberalism amounts to writing intellectual history on the basis of verbal coincidences. Ditto for the relation between the axiom of choice[115] and the movement for abortion rights. Cohen has indeed shown that neither the axiom of choice nor its negation can be deduced from the other axioms of set theory; but this mathematical result has no political implications whatsoever.

Finally, all the bibliographic entries are rigorously exact, apart from a wink at former French minister of culture Jacques Toubon, who tried to impose the use of French in scientific conferences sponsored by the French government (see Kontsevitch 1994), and at Catalan nationalism (see Smolin 1992).

---

[115]See p. 44 above for a brief explanation of the axiom of choice.

C. Transgressing the Boundaries: An Afterword*

Les grandes personnes sont décidément bien bizarres, se dit le petit prince.

—*Antoine de Saint Exupéry, Le Petit Prince*

Alas, the truth is out: my article, "Transgressing the Boundaries: Toward a Transformative Hermeneutics of Quantum Gravity", which appeared in the spring/summer 1996 issue of the cultural-studies journal *Social Text*, is a parody. Clearly I owe the editors and readers of *Social Text*, as well as the wider intellectual community, a non-parodic explanation of my motives and my true views.[1] One of my goals here is to make a small contribution toward a dialogue on the Left between humanists and natural scientists—"two cultures" which, contrary to some optimistic pronouncements (mostly by the former group), are probably farther apart in mentality than at any time in the past fifty years.

Like the genre it is meant to satirize—myriad exemplars of which can be found in my reference list—my article is a mélange of truths, half-truths, quarter-truths, falsehoods, non

---

*This article was submitted to *Social Text* following the publication of the parody, but was rejected on the grounds that it did not meet their intellectual standards. It was published in *Dissent* **43**(4), pp. 93–99 (Fall 1996) and, in slightly different form, in *Philosophy and Literature* **20**(2), pp. 338–346 (October 1996). See also the critical comment by *Social Text* co-founder Stanley Aronowitz (1997) and the reply by Sokal (1997b).

[1] Readers are cautioned not to infer my views on any subject except insofar as they are set forth in this Afterword. In particular, the fact that I have parodied an extreme or ambiguously stated version of an idea does not exclude that I may agree with a more nuanced or precisely stated version of the same idea.

sequiturs, and syntactically correct sentences that have no meaning whatsoever. (Sadly, there are only a handful of the latter: I tried hard to produce them, but I found that, save for rare bursts of inspiration, I just didn't have the knack.) I also employed some other strategies that are well-established (albeit sometimes inadvertently) in the genre: appeals to authority in lieu of logic; speculative theories passed off as established science; strained and even absurd analogies; rhetoric that sounds good but whose meaning is ambiguous; and confusion between the technical and everyday senses of English words.[2] (N.B. All works cited in my article are real, and all quotations are rigorously accurate; none are invented.)

But why did I do it? I confess that I'm an unabashed Old Leftist who never quite understood how deconstruction was supposed to help the working class. And I'm a stodgy old scientist who believes, naively, that there exists an external world, that there exist objective truths about that world, and that my job is to discover some of them. (If science were merely a negotiation of social conventions about what is agreed to be "true", why would I bother devoting a large fraction of my all-too-short life to it? I don't aspire to be the Emily Post of quantum field theory.[3])

But my main concern isn't to defend science from the barbarian hordes of lit crit (we'll survive just fine, thank you). Rather, my concern is explicitly *political*: to combat a currently fashionable postmodernist/poststructuralist/social-constructivist discourse—and more generally a penchant for subjectivism—

---

[2]For example: "linear", "nonlinear", "local", "global", "multidimensional", "relative", "frame of reference", "field", "anomaly", "chaos", "catastrophe", "logic", "irrational", "imaginary", "complex", "real", "equality", "choice".

[3]By the way, anyone who believes that the laws of physics are mere social conventions is invited to try transgressing those conventions from the windows of my apartment. I live on the twenty-first floor. (P.S. I am aware that this wisecrack is unfair to the more sophisticated relativist philosophers of science, who will concede that *empirical statements* can be objectively true—e.g. the fall from my window to the pavement will take approximately 2.5 seconds—but claim that the *theoretical explanations* of those empirical statements are more-or-less arbitrary social constructions. I think that also this view is largely wrong, but that is a much longer discussion.)

which is, I believe, inimical to the values and future of the Left.[4]
Alan Ryan said it well:

> It is, for instance, pretty suicidal for embattled minorities to
> embrace Michel Foucault, let alone Jacques Derrida. The mi-
> nority view was always that power could be undermined by
> truth . . . Once you read Foucault as saying that truth is sim-
> ply an effect of power, you've had it. . . . But American de-
> partments of literature, history and sociology contain large
> numbers of self-described leftists who have confused radical
> doubts about objectivity with political radicalism, and are in
> a mess.[5]

Likewise, Eric Hobsbawm has decried

> the rise of "postmodernist" intellectual fashions in Western
> universities, particularly in departments of literature and an-
> thropology, which imply that all "facts" claiming objective ex-
> istence are simply intellectual constructions. In short, that
> there is no clear difference between fact and fiction. But there

---

[4]The natural sciences have little to fear, at least in the short run, from postmodernist
silliness; it is, above all, history and the social sciences—and leftist politics—that
suffer when verbal game-playing displaces the rigorous analysis of social realities.
Nevertheless, because of the limitations of my own expertise, my analysis here will
be restricted to the natural sciences (and indeed primarily to the physical sciences).
While the basic epistemology of inquiry ought to be roughly the same for the natural
and social sciences, I am of course perfectly aware that many special (and very
difficult) methodological issues arise in the social sciences from the fact that the
objects of inquiry are human beings (including their subjective states of mind); that
these objects of inquiry have intentions (including in some cases the concealment of
evidence or the placement of deliberately self-serving evidence); that the evidence is
expressed (usually) in human language whose meaning may be ambiguous; that the
meaning of conceptual categories (e.g. childhood, masculinity, femininity, family,
economics, etc.) changes over time; that the goal of historical inquiry is not just facts
but interpretation, etc. So by no means do I claim that my comments about physics
should apply directly to history and the social sciences—that would be absurd. To
say that "physical reality is a social and linguistic construct" is just plain silly, but to
say that "social reality is a social and linguistic construct" is virtually a tautology.

[5]Ryan (1992).

is, and for historians, even for the most militantly antipositivist ones among us, the ability to distinguish between the two is absolutely fundamental.[6]

(Hobsbawm goes on to show how rigorous historical work can refute the fictions propounded by reactionary nationalists in India, Israel, the Balkans and elsewhere.) And finally Stanislav Andreski:

> So long as authority inspires awe, confusion and absurdity enhance conservative tendencies in society. Firstly, because clear and logical thinking leads to a cumulation of knowledge (of which the progress of the natural sciences provides the best example) and the advance of knowledge sooner or later undermines the traditional order. Confused thinking, on the other hand, leads nowhere in particular and can be indulged indefinitely without producing any impact upon the world.[7]

As an example of "confused thinking", I would like to consider a chapter from Harding (1991) entitled "Why 'Physics' Is a Bad Model for Physics". I select this example both because of Harding's prestige in certain (but by no means all) feminist circles, and because her essay is (unlike much of this genre) very clearly written. Harding wishes to answer the question, "Are feminist criticisms of Western thought relevant to the natural sciences?" She does so by raising, and then rebutting, six "false beliefs" about the nature of science. Some of her rebuttals are perfectly well-taken; but they don't prove anything like what she claims they do. That is because she conflates five quite distinct issues:

---

[6]Hobsbawm (1993, p. 63).

[7]Andreski (1972, p. 90).

1) *Ontology.* What objects *exist* in the world? What statements about these objects are *true*?

2) *Epistemology.* How can human beings obtain *knowledge* of truths about the world? How can they assess the *reliability* of that knowledge?

3) *Sociology of knowledge.* To what extent are the truths *known* (or *knowable*) by humans in any given society influenced (or determined) by social, economic, political, cultural, and ideological factors? Same question for the false statements erroneously believed to be true.

4) *Individual ethics.* What types of research *ought* a scientist (or technologist) to undertake (or refuse to undertake)?

5) *Social ethics.* What types of research *ought* society to encourage, subsidize, or publicly fund (or alternatively to discourage, tax, or forbid)?

These questions are obviously related—e.g. if there are no objective truths about the world, then there isn't much point in asking how one can know those (nonexistent) truths—but they are conceptually distinct.

For example, Harding (citing Forman 1987) points out that American research in the 1940s and 50s on quantum electronics was motivated in large part by potential military applications. True enough. Now, quantum mechanics made possible solid-state physics, which in turn made possible quantum electronics (e.g. the transistor), which made possible nearly all of modern technology (e.g. the computer).[8] And the computer has had applications that are beneficial to society (e.g. in allowing the postmodern cultural critic to produce her articles more efficiently)

---

[8]Computers existed prior to solid-state technology, but they were unwieldy and slow. The 486 PC sitting today on the literary theorist's desk is roughly 1000 times more powerful than the room-sized vacuum-tube computer IBM 704 from 1954 (see e.g. Williams 1985).

as well as applications that are harmful (e.g. in allowing the U.S. military to kill human beings more efficiently). This raises a host of social and individual ethical questions: Ought society to forbid (or discourage) certain applications of computers? Forbid (or discourage) research on computers *per se*? Forbid (or discourage) research on quantum electronics? On solid-state physics? On quantum mechanics? And likewise for individual scientists and technologists. (Clearly, an affirmative answer to these questions becomes harder to justify as one goes down the list; but I do not want to declare any of these questions *a priori* illegitimate.) Likewise, sociological questions arise, for example: To what extent is our (true) knowledge of computer science, quantum electronics, solid-state physics, and quantum mechanics—and our lack of knowledge about other scientific subjects, e.g. the global climate—a result of public-policy choices favoring militarism? To what extent have the erroneous theories (if any) in computer science, quantum electronics, solid-state physics, and quantum mechanics been the result (in whole or in part) of social, economic, political, cultural, and ideological factors, in particular the culture of militarism?[9] These are all serious questions, which deserve careful investigation adhering to the highest standards of scientific and historical evidence. *But they have no effect whatsoever on the underlying scientific questions:* whether atoms (and silicon crystals, transistors, and computers) really do behave according to the laws of quantum mechanics (and solid-state physics, quantum electronics, and computer science). The militaristic orientation of American science has quite simply no bearing whatsoever on the ontological question, and only under a wildly implausible scenario could it have any bearing on the epistemological question. (E.g. if the worldwide community of solid-

---

[9]I certainly don't exclude the possibility that *present* theories in any of these subjects might be erroneous. But critics wishing to make such a case would have to provide not only historical evidence of the claimed cultural influence, but also *scientific* evidence that the theory in question is in fact erroneous. (The same evidentiary standards of course apply to *past* erroneous theories; but in this case the scientists may have already performed the second task, relieving the cultural critic of the need to do so from scratch.)

state physicists, following what they believe to be the conventional standards of scientific evidence, were to hastily accept an erroneous theory of semiconductor behavior because of their enthusiasm for the breakthrough in military technology that this theory would make possible.)

Andrew Ross has drawn an analogy between the hierarchical taste cultures (high, middlebrow, and popular) familiar to cultural critics, and the demarcation between science and pseudoscience.[10] At a sociological level this is an incisive observation; but at an ontological and epistemological level it is simply mad. Ross seems to recognize this, because he immediately says:

> I do not want to insist on a literal interpretation of this analogy . . . A more exhaustive treatment would take account of the local, qualifying differences between the realm of cultural taste and that of science [!], but it would run up, finally, against the stand-off between the empiricist's claim that non-context-dependent beliefs exist and that they can be true, and the culturalist's claim that beliefs are only socially accepted as true.[11]

But such epistemological agnosticism simply won't suffice, at least not for people who aspire to make social change. Deny that non-context-dependent assertions can be true, and you don't just throw out quantum mechanics and molecular biology: you also throw out the Nazi gas chambers, the American

---

[10]Ross (1991, p. 25–26); also in Ross (1992, pp. 535–536).

[11]Ross (1991, p. 26); also in Ross (1992, p. 535). In the discussion following this paper, Ross (1992, p. 549) expressed further (and quite justified) misgivings:

> I'm quite skeptical of the "anything goes" spirit that is often the prevailing climate of relativism around postmodernism. . . . Much of the postmodernist debate has been devoted to grappling with the philosophical or cultural limits to the grand narratives of the Enlightenment. If you think about ecological questions in this light, however, then you are talking about "real" physical, or material, limits to our resources for encouraging social growth. And postmodernism, as we know, has been loath to address the "real," except to announce its banishment.

enslavement of Africans, and the fact that today in New York it's raining. Hobsbawm is right: facts do matter, and some facts (like the first two cited here) matter a great deal.

Still, Ross is correct that, at a sociological level, maintaining the demarcation line between science and pseudoscience serves—*among other things*—to maintain the social power of those who, whether or not they have formal scientific credentials, stand on science's side of the line. (It has *also* served to increase the mean life expectancy in the United States from 47 years to 76 years in less than a century.[12]) Ross notes that

> Cultural critics have, for some time now, been faced with the task of exposing similar vested institutional interests in the debates about class, gender, race, and sexual preference that touch upon the demarcations between taste cultures, and I see no ultimate reason for us to abandon our hard-earned skepticism when we confront science.[13]

Fair enough: scientists are in fact the *first* to advise skepticism in the face of other people's (and one's own) truth claims. But a sophomoric skepticism, a bland (or blind) agnosticism, won't get you anywhere. Cultural critics, like historians or scientists,

---

[12]U.S. Bureau of the Census (1975, pp. 47, 55; 1994, p. 87). In 1900 the mean life expectancy at birth was 47.3 years (47.6 years for whites, and a shocking 33.0 years for "Negro and other"). In 1995 it is 76.3 years (77.0 years for whites, 70.3 years for blacks).

I am aware that this assertion is likely to be misinterpreted, so let me engage in some pre-emptive clarification. I am *not* claiming that all of the increase in life expectancy is due to advances in scientific *medicine*. A large fraction (possibly the dominant part) of the increase—especially in the first three decades of the twentieth century—is due to the general improvement in the standards of housing, nutrition, and public sanitation (the latter two informed by improved scientific understanding of the etiology of infectious and dietary-deficiency diseases). [For reviews of the evidence, see e.g. Holland *et al.* (1991).] But—without discounting the role of social struggles in these improvements, particularly as concerns the narrowing of the racial gap—the underlying and overwhelming cause of these improvements is quite obviously the vast increase in the material standard of living over the past century, by more than a factor of five (U.S. Bureau of the Census 1975, pp. 224–225; 1994, p. 451). And this increase is quite obviously the direct result of science, as embodied in technology.

[13]Ross (1991, p. 26); also in Ross (1992, p. 536).

need an *informed* skepticism: one that can evaluate evidence and logic, and come to reasoned (albeit tentative) judgments *based on that evidence and logic.*

At this point Ross may object that I am rigging the power game in my own favor: how is he, a professor of American Studies, to compete with me, a physicist, in a discussion of quantum mechanics?[14] (Or even of nuclear power—a subject on which I have no expertise whatsoever.) But it is equally true that I would be unlikely to win a debate with a professional historian on the causes of World War I. Nevertheless, as an intelligent lay person with a modest knowledge of history, I am capable of evaluating the evidence and logic offered by competing historians, and of coming to some sort of reasoned (albeit tentative) judgment. (Without that ability, how could any thoughtful person justify being politically active?)

The trouble is that few non-scientists in our society feel this self-confidence when dealing with scientific matters. As C.P. Snow observed in his famous "Two Cultures" lecture 35 years ago:

> A good many times I have been present at gatherings of people who, by the standards of the traditional culture, are thought highly educated and who have with considerable gusto been expressing their incredulity at the illiteracy of scientists. Once or twice I have been provoked and have asked the company how many of them could describe the Second Law of Thermodynamics. The response was cold: it was also negative. Yet I was asking something which is about the scientific equivalent of: *Have you read a work of Shakespeare's?*

---

[14]By the way, intelligent non-scientists seriously interested in the conceptual problems raised by quantum mechanics need no longer rely on the vulgarizations (in both senses) published by Heisenberg, Bohr, and sundry physicists and New Age authors. The little book of Albert (1992) provides an impressively serious and *intellectually honest* account of quantum mechanics and the philosophical issues it raises—yet it requires no more mathematical background than a modicum of high-school algebra, and does not require any prior knowledge of physics. The main requirement is a willingness to think *slowly* and *clearly*.

I now believe that if I had asked an even simpler question—such as, What do you mean by mass, or acceleration, which is the scientific equivalent of saying, *Can you read?*—not more than one in ten of the highly educated would have felt that I was speaking the same language. So the great edifice of modern physics goes up, and the majority of the cleverest people in the western world have about as much insight into it as their neolithic ancestors would have had.[15]

A lot of the blame for this state of affairs rests, I think, with the scientists. The teaching of mathematics and science *is* often authoritarian[16]; and this is antithetical not only to the principles of radical/democratic pedagogy but to the principles of science itself. No wonder most Americans can't distinguish between science and pseudoscience: their science teachers have never given them any rational grounds for doing so. (Ask an average undergraduate: Is matter composed of atoms? Yes. Why do you think so? The reader can fill in the response.) Is it then any surprise that 36% of Americans believe in telepathy, and that 47% believe in the creation account of Genesis?[17]

---

[15]Snow (1963, pp. 20–21). One significant change has taken place since C.P. Snow's time: while humanist intellectuals' ignorance about (for example) mass and acceleration remains substantially unchanged, nowadays a significant minority of humanist intellectuals feels entitled to pontificate on these subjects in spite of their ignorance (perhaps trusting that their readers will be equally ignorant). Consider, for example, the following excerpt from a recent book on *Rethinking Technologies*, edited by the Miami Theory Collective and published by the University of Minnesota Press: "it now seems appropriate to reconsider the notions of acceleration and deceleration (what physicists call positive and negative speeds)" (Virilio 1993, p. 5). The reader who does not find this uproariously funny (as well as depressing) is invited to sit in on the first two weeks of Physics I.

[16]I wasn't joking about that. For anyone who is interested in my views, I would be glad to provide a copy of Sokal (1987). For another sharp critique of the poor teaching of mathematics and science, see (irony of ironies) Gross and Levitt (1994, pp. 23–28).

[17]Telepathy: Hastings and Hastings (1992, p. 518), American Institute of Public Opinion poll from June 1990. Concerning "telepathy, or communication between minds without using the traditional five senses", 36% "believe in", 25% are "not sure", and 39% "do not believe in". For "people on this earth are sometimes possessed by the devil", it is 49–16–35 (!). For "astrology, or that the position of the stars and planets can affect people's lives", it is 25–22–53. Mercifully, only 11% believe in channeling (22% are not sure), and 7% in the healing power of pyramids (26% not sure).

As Ross has noted[18], many of the central political issues of the coming decades—from health care to global warming to Third World development—depend in part on subtle (and hotly debated) questions of scientific fact. But they don't depend only on scientific fact: they depend also on ethical values and—in this journal it hardly needs to be added—on naked economic interests. No Left can be effective unless it takes seriously questions of scientific fact *and* of ethical values *and* of economic interests. The issues at stake are too important to be left to the capitalists or to the scientists—or to the postmodernists.

A quarter-century ago, at the height of the U.S. invasion of Vietnam, Noam Chomsky observed that

> George Orwell once remarked that political thought, especially on the left, is a sort of masturbation fantasy in which the world of fact hardly matters. That's true, unfortunately, and it's part of the reason that our society lacks a genuine, responsible, serious left-wing movement.[19]

Perhaps that's unduly harsh, but there's unfortunately a significant kernel of truth in it. Nowadays the erotic text tends to be written in (broken) French rather than Chinese, but the real-life consequences remain the same. Here's Alan Ryan in 1992, con-

---

Creationism: Gallup (1993, pp. 157–159), Gallup poll from June 1993. The exact question was: "Which of the following statements comes closest to your views on the origin and development of human beings: 1) human beings have developed over millions of years from less advanced forms of life, but God guided this process; 2) human beings have developed over millions of years from less advanced forms of life, but God had no part in this process; 3) God created human beings pretty much in their present form at one time within the last 10,000 years or so?" The results were 35% developed with God, 11% developed without God, 47% God created in present form, 7% no opinion. A poll from July 1982 (Gallup 1982, pp. 208–214) found almost identical figures, but gave breakdowns by sex, race, education, region, age, income, religion, and community size. Differences by sex, race, region, income, and (surprisingly) religion were rather small. By far the largest difference was by education: only 24% of college graduates supported creationism, compared to 49% of high-school graduates and 52% of those with a grade-school education. So maybe the worst science teaching is at the elementary and secondary levels.

[18]See Note 11 above.

[19]Chomsky (1984, p. 200), lecture delivered in 1969.

cluding his wry analysis of American intellectual fashions with a lament that

> the number of people who combine intellectual toughness with even a modest political radicalism is pitifully small. Which, in a country that has George Bush as President and Danforth Quayle lined up for 1996, is not very funny.[20]

Four years later, with Bill Clinton installed as our supposedly "progressive" president and Newt Gingrich already preparing for the new millennium, it still isn't funny.

Works Cited

Albert, David Z. 1992. *Quantum Mechanics and Experience*. Cambridge: Harvard University Press.

Andreski, Stanislav. 1972. *Social Sciences as Sorcery*. London: André Deutsch.

Chomsky, Noam. 1984. The politicization of the university. In *Radical Priorities*, 2nd ed., pp. 189–206, edited by Carlos P. Otero. Montreal: Black Rose Books.

Forman, Paul. 1987. Behind quantum electronics: National security as basis for physical research in the United States, 1940–1960. *Historical Studies in the Physical and Biological Sciences* **18**: 149–229.

Gallup, George H. 1982. *The Gallup Poll: Public Opinion 1982*. Wilmington, Del.: Scholarly Resources.

Gallup, George Jr. 1993. *The Gallup Poll: Public Opinion 1993*. Wilmington, Del.: Scholarly Resources.

Gross, Paul R. and Norman Levitt. 1994. The natural sciences: Trouble ahead? Yes. *Academic Questions* **7**(2): 13–29.

Harding, Sandra. 1991. *Whose Science? Whose Knowledge? Thinking from Women's Lives*. Ithaca, N.Y.: Cornell University Press.

Hastings, Elizabeth Hann and Philip K. Hastings, eds. 1992. *Index to International Public Opinion, 1990–1991*. New York: Greenwood Press.

Hobsbawm, Eric. 1993. The new threat to history. *New York Review of Books* (16 December): 62–64.

---

[20]Ryan (1992).

Holland, Walter W. *et al.*, eds. 1991. *Oxford Textbook of Public Health*, 3 vols. Oxford: Oxford University Press.

Ross, Andrew. 1991. *Strange Weather: Culture, Science, and Technology in the Age of Limits*. London: Verso.

Ross, Andrew. 1992. New Age technocultures. In *Cultural Studies*, pp. 531–555, edited by Lawrence Grossberg, Cary Nelson and Paula A. Treichler. New York: Routledge.

Ryan, Alan. 1992. Princeton diary. *London Review of Books* (26 March): 21.

Snow, C.P. 1963. *The Two Cultures: And A Second Look*. New York: Cambridge University Press.

Sokal, Alan. 1987. Informe sobre el plan de estudios de las carreras de Matemática, Estadística y Computación. Report to the Universidad Nacional Autónoma de Nicaragua, Managua, unpublished.

U.S. Bureau of the Census. 1975. *Historical Statistics of the United States: Colonial Times to 1970*. Washington: Government Printing Office.

U.S. Bureau of the Census. 1994. *Statistical Abstract of the United States: 1994*. Washington: Government Printing Office.

Virilio, Paul. 1993. The third interval: A critical transition. In *Rethinking Technologies*, pp. 3–12, edited by Verena Andermatt Conley on behalf of the Miami Theory Collective. Minneapolis: University of Minnesota Press.

Williams, Michael R. 1985. *A History of Computing Technology*. Englewood Cliffs, N.J.: Prentice-Hall.

Bibliography

Albert, David Z. 1992. *Quantum Mechanics and Experience*. Cambridge, Massachusetts: Harvard University Press.

Albert, Michael. 1992–93. "Not all stories are equal: Michael Albert answers the pomo advocates". *Z Papers Special Issue on Postmodernism and Rationality*. Available on-line at http://www.zmag.org/zmag/articles/albertpomoreply.html

Albert, Michael. 1996. "Science, post modernism and the left". *Z Magazine* **9**(7/8) (July/August): 64–69.

Alliez, Eric. 1993. *La Signature du monde, ou Qu'est-ce que la philosophie de Deleuze et Guattari?* Paris: Éditions du Cerf.

Althusser, Louis. 1993. *Écrits sur la psychanalyse: Freud et Lacan*. Paris: Stock/IMEC.

Amsterdamska, Olga. 1990. "Surely you are joking, Monsieur Latour!" *Science, Technology, & Human Values* **15**: 495–504.

Andreski, Stanislav. 1972. *Social Sciences as Sorcery*. London: André Deutsch.

Anyon, Roger, T.J. Ferguson, Loretta Jackson and Lillie Lane. 1996. "Native American oral traditions and archaeology". *SAA Bulletin* [Bulletin of the Society for American Archaeology] **14**(2) (March/April): 14–16. Available on-line at http://www.sscf.ucsb.edu/SAABulletin/14.2/SAA14.html

Arnol'd, Vladimir I. 1992. *Catastrophe Theory*, 3rd ed. Translated by G.S. Wassermann and R.K. Thomas. Berlin: Springer-Verlag.

Aronowitz, Stanley. 1997. "Alan Sokal's 'Transgression' ". *Dissent* **44**(1) (Winter): 107–110.

Badiou, Alain. 1982. *Théorie du sujet*. Paris: Seuil.

Bahcall, John N. 1990. "The solar-neutrino problem". *Scientific American* **262**(5) (May): 54–61.

Bahcall, John N., Frank Calaprice, Arthur B. McDonald, and Yoji Totsuka. 1996. "Solar neutrino experiments: The next generation". *Physics Today* **49**(7) (July): 30–36.

Barnes, Barry and David Bloor. 1981. "Relativism, rationalism and the sociology of knowledge". In *Rationality and Relativism*, pages 21–47. Edited by Martin Hollis and Steven Lukes. Oxford: Blackwell.

Barnes, Barry, David Bloor, and John Henry. 1996. *Scientific Knowledge: A Sociological Analysis*. Chicago: University of Chicago Press.

Barsky, Robert F. 1997. *Noam Chomsky: A Life of Dissent*. Cambridge, Massachusetts: MIT Press.

Barthes, Roland. 1970. "L'étrangère". *La Quinzaine Littéraire* **94** (1–15 mai 1970): 19–20.

Baudrillard, Jean. 1990. *Fatal Strategies*. Translated by Philip Beitchman and W.G.J. Niesluchowski. Edited by Jim Fleming. New York: Semiotext(e). [French original: *Les Stratégies fatales*. Paris: Bernard Grasset, 1983.]

Baudrillard, Jean. 1993. *The Transparency of Evil: Essays on Extreme Phenomena*. Translated by James Benedict. London: Verso. [French original: *La Transparence du mal*. Paris: Galilée, 1990.]

Baudrillard, Jean. 1994. *The Illusion of the End*. Translated by Chris Turner. Cambridge, England: Polity Press. [French original: *L'Illusion de la fin*. Paris: Galilée, 1992.]

Baudrillard, Jean. 1995. *The Gulf War Did Not Take Place*. Translated and with an introduction by Paul Patton. Bloomington: Indiana University Press. [French original: *La Guerre du Golfe n'a pas eu lieu*. Paris: Galilée, 1991.]

Baudrillard, Jean. 1996. *The Perfect Crime*. Translated by Chris Turner. London: Verso. [French original: *Le Crime parfait*. Paris: Galilée, 1995.]

Baudrillard, Jean. 1997. *Fragments: Cool Memories III, 1990–1995*. Translated by Emily Agar. London: Verso. [French original: *Fragments; Cool memories III 1990–1995*. Paris: Galilée, 1995.]

Best, Steven. 1991. "Chaos and entropy: Metaphors in postmodern science and social theory." *Science as Culture* **2**(2) (no. 11): 188–226.

Bloor, David. 1991. *Knowledge and Social Imagery*. 2nd ed. Chicago: University of Chicago Press.

Boghossian, Paul. 1996. "What the Sokal hoax ought to teach us." *Times Literary Supplement* (13 December): 14–15.

Bouveresse, Jacques. 1984. *Rationalité et cynisme*. Paris: Éditions de Minuit.

Boyer, Carl B. 1959 [1949]. *The History of the Calculus and Its Conceptual Development*. With a foreword by R. Courant. New York: Dover.

Brecht, Bertolt. 1965. *The Messingkauf Dialogues*. Translated by John Willett. London: Methuen.

Bricmont, Jean. 1995a. "Science of chaos or chaos in science?" *Physicalia Magazine* **17**, no. 3–4. Available on-line as publication UCL-IPT-96-03 at http://www.fyma.ucl.ac.be/reche/1996/1996.html [A slightly earlier version of this article appeared in Paul R. Gross, Norman Levitt and Martin W. Lewis, eds., *The Flight from Science and Reason, Annals of the New York Academy of Sciences* **775** (1996), pp. 131–175.]

Bricmont, Jean. 1995b. "Contre la philosophie de la mécanique quantique". In *Les Sciences et la philosophie. Quatorze essais de rapprochement*, pp. 131–179. Edited by R. Franck. Paris: Vrin.

Broch, Henri. 1992. *Au Cœur de l'extraordinaire*. Bordeaux: L'Horizon Chimérique.

Brunet, Pierre. 1931. *L'Introduction des théories de Newton en France au XVIIIᵉ siècle*. Paris: A. Blanchard. Reprinted by Slatkine, Genève, 1970.

Brush, Stephen. 1989. "Prediction and theory evaluation: The case of light bending". *Science* **246**: 1124–1129.

Canning, Peter. 1994. "The crack of time and the ideal game". In *Gilles Deleuze and the Theater of Philosophy*, pp. 73–98. Edited by Constantin V. Boundas and Dorothea Olkowski. New York: Routledge.

Chomsky, Noam. 1979. *Language and Responsibility*. Based on conversations with Mitsou Ronat. Translated by John Viertel. New York: Pantheon. [French original: *Dialogues avec Mitsou Ronat*. Paris: Flammarion, 1977.]

Chomsky, Noam. 1992–93. "Rationality/Science". *Z Papers Special Issue on Postmodernism and Rationality*. Available on-line at http://www.zmag.org/zmag/articles/chompomoart.html

Chomsky, Noam. 1993. *Year 501: The Conquest Continues*. Boston: South End Press.

Chomsky, Noam. 1994. *Keeping the Rabble in Line: Interviews with David Barsamian*. Monroe, Maine: Common Courage Press.

Clavelin, Maurice. 1994. "L'histoire des sciences devant la sociologie de la science". In *Le Relativisme est-il résistible? Regards sur la sociologie des sciences*, pp. 229–247. Edited by Raymond Boudon and Maurice Clavelin. Paris: Presses Universitaires de France.

Coutty, Marc. 1998. "Des normaliens jugent l'affaire Sokal". Interview with Mikaël Cozic, Grégoire Kantardjian and Léon Loiseau. *Le Monde de l'Éducation* **255** (January): 8–10.

Crane, H.R. 1968. "The *g* factor of the electron". *Scientific American* **218**(1) (January): 72–85.

Dahan-Dalmedico, Amy. 1997. "Rire ou frémir?" *La Recherche* **304** (December): 10. [An extended version of this article appears in *Revue de l'Association Henri Poincaré* **9**(7), décembre 1997, pp. 15–18.]

Damarin, Suzanne K. 1995. "Gender and mathematics from a feminist standpoint". In: *New Directions for Equity in Mathematics Education*, pp. 242–257. Edited by Walter G. Secada, Elizabeth Fennema and Lisa Byrd Adajian. Published in collaboration with the National Council of Teachers of Mathematics. New York: Cambridge University Press.

Darmon, Marc. 1990. *Essais sur la topologie lacanienne*. Paris: Éditions de l'Assocation Freudienne.

Davenas, E. *et al.* 1988. "Human basophil degranulation triggered by very dilute antiserum against IgE". *Nature* **333**: 816–818.

Davis, Donald M. 1993. *The Nature and Power of Mathematics*. Princeton: Princeton University Press.

Dawkins, Richard. 1986. *The Blind Watchmaker.* New York: Norton.

Debray, Régis. 1980. *Le Scribe: Genèse du politique.* Paris: Bernard Grasset.

Debray, Régis. 1981. *Critique de la raison politique.* Paris: Gallimard.

Debray, Régis. 1983. *Critique of Political Reason.* Translated by David Macey. London: New Left Books. [French original: see Debray 1981.]

Debray, Régis. 1994. *Manifestes médiologiques.* Paris: Gallimard.

Debray, Régis. 1996a. *Media Manifestos: On the Technological Transmission of Cultural Forms.* Translated by Eric Rauth. London: Verso. [French original: see Debray 1994.]

Debray, Régis. 1996b. "L'incomplétude, logique du religieux?" *Bulletin de la société française de philosophie* **90** (session of 27 January): 1–35.

de Gennes, Pierre-Gilles. 1976. "La percolation: un concept unificateur". *La Recherche* **72**: 919–927.

Deleuze, Gilles. 1990. *The Logic of Sense.* Translated by Mark Lester with Charles Stivale. Edited by Constantin V. Boundas. New York: Columbia University Press. [French original: *Logique du sens.* Paris: Éditions de Minuit, 1969.]

Deleuze, Gilles. 1994. *Difference and Repetition.* Translated by Paul Patton. New York: Columbia University Press. [French original: *Différence et répétition.* Paris: Presses Universitaires de France, 1968.]

Deleuze, Gilles, and Félix Guattari. 1987. *A Thousand Plateaus: Capitalism and Schizophrenia.* Translation and foreword by Brian Massumi. Minneapolis: University of Minnesota Press. [French original: *Mille plateaux.* Paris: Éditions de Minuit, 1980.]

Deleuze, Gilles and Félix Guattari. 1994. *What is Philosophy?* Translated by Hugh Tomlinson and Graham Burchell. New York: Columbia University Press. [French original: *Qu'est-ce que la philosophie?* Paris: Éditions de Minuit, 1991.]

Derrida, Jacques. 1970. "Structure, sign and play in the discourse of the human sciences". In *The Languages of Criticism and the Sciences of Man: The Structuralist Controversy*, pp. 247–272. Edited by Richard Macksey and Eugenio Donato. Baltimore: Johns Hopkins Press.

Desanti, Jean Toussaint. 1975. *La Philosophie silencieuse, ou critique des philosophies de la science.* Paris: Éditions du Seuil.

Devitt, Michael. 1997. *Realism and Truth*, 2nd edition with a new afterword. Princeton: Princeton University Press.

Dhombres, Jean. 1994. "L'histoire des sciences mise en question par les approches sociologiques: le cas de la communauté scientifique française (1789–1815)". In *Le Relativisme est-il résistible? Regards sur la sociologie des sciences*, pp. 159–205. Edited by Raymond Boudon and Maurice Clavelin. Paris: Presses Universitaires de France.

Dieudonné, Jean Alexandre. 1989. *A History of Algebraic and Differential Topology, 1900–1960.* Boston: Birkhäuser.

Dobbs, Betty Jo Teeter, and Margaret C. Jacob. 1995. *Newton and the Culture of Newtonianism.* Atlantic Highlands, New Jersey: Humanities Press.

Donovan, Arthur, Larry Laudan, and Rachel Laudan. 1988. *Scrutinizing Science: Empirical Studies of Scientific Change*. Dordrecht–Boston: Kluwer Academic Publishers.

Droit, Roger-Pol. 1997. "Au risque du 'scientifiquement correct' ". *Le Monde* (30 September): 27.

D'Souza, Dinesh. 1991. *Illiberal Education: The Politics of Race and Sex on Campus*. New York: Free Press.

Duhem, Pierre. 1954 [1914]. *The Aim and Structure of Physical Theory*. Translated by Philip P. Wiener. Princeton: Princeton University Press. [French original: *La Théorie physique: son objet, sa structure*, 2ème éd. revue et augmentée. Paris: Rivière, 1914.]

Dumm, Thomas, Anne Norton *et al.* 1998. "On left conservatism". Proceedings of a workshop at the University of California–Santa Cruz, 31 January 1998. *Theory & Event*, issues 2.2 and 2.3. Available on-line at http://muse.jhu.edu/journals/theory_&_event/

Eagleton, Terry. 1995. "Where do postmodernists come from?" *Monthly Review* **47**(3) (July/August): 59–70. [Reprinted in Ellen Meiksins Wood and John Bellamy Foster, eds., *In Defense of History*, New York: Monthly Review Press, 1997, pp. 17–25; and in Terry Eagleton, *The Illusions of Postmodernism*, Oxford: Blackwell, 1996.]

Economist (unsigned). 1997. "You can't follow the science wars without a battle map". *The Economist* (13 December): 77–79.

Ehrenreich, Barbara. 1992–93. "For the rationality debate". *Z Papers Special Issue on Postmodernism and Rationality*. Available on-line at http://www.zmag.org/zmag/articles/ehrenrationpiece.html

Einstein, Albert. 1949. "Remarks concerning the essays brought together in this cooperative volume". In *Albert Einstein, Philosopher-Scientist*, pages 665–688. Edited by Paul Arthur Schilpp. Evanston, Illinois: Library of Living Philosophers.

Einstein, Albert. 1960 [1920]. *Relativity: The Special and the General Theory*. London: Methuen.

Epstein, Barbara. 1995. "Why poststructuralism is a dead end for progressive thought". *Socialist Review* **25**(2): 83–120.

Epstein, Barbara. 1997. "Postmodernism and the left". *New Politics* **6**(2) (Winter): 130–144.

Eribon, Didier. 1994. *Michel Foucault et ses contemporains*. Paris: Fayard.

Euler, Leonhard. 1997 [1761]. "Refutation of the idealists". In *Letters of Euler to a German Princess*, volume 1, letter XCVII, pp. 426–430. Translated by Henry Hunter (originally published 1795). With a new introduction by Andrew Pyle. London: Thoemmes Press. [French original: Lettres à une princesse d'Allemagne, lettre 97. In: *Leonhardi Euleri Opera Omnia*, série III, volume 11, pp. 219–220. Turici, 1911–.]

Ferguson, Euan. 1996. "Illogical dons swallow hoaxer's quantum leap into gibberish". *The Observer* [London] (19 May): 1.

Feyerabend, Paul. 1975. *Against Method*. London: New Left Books.

Feyerabend, Paul. 1987. *Farewell to Reason*. London: Verso.

Feyerabend, Paul. 1988. *Against Method*, 2nd ed. London: Verso.

Feyerabend, Paul. 1992. "Atoms and consciousness". *Common Knowledge* 1(1): 28–32.

Feyerabend, Paul. 1993. *Against Method*, 3rd ed. London: Verso.

Feyerabend, Paul. 1995. *Killing Time: The Autobiography of Paul Feyerabend*. Chicago: University of Chicago Press.

Feynman, Richard. 1965. *The Character of Physical Law*. Cambridge, Massachusetts: MIT Press.

Foucault, Michel. 1970. "Theatrum philosophicum". *Critique* **282**: 885–908.

Fourez, Gérard. 1992. *La Construction des sciences*, 2ème édition revue. Brussels: De Boeck Université.

Fourez, Gérard, Véronique Englebert-Lecomte, and Philippe Mathy. 1997. *Nos savoirs sur nos savoirs: Un lexique d'épistémologie pour l'enseignement*. Brussels: De Boeck Université.

Frank, Tom. 1996. "Textual reckoning". *In These Times* **20**(14) (27 May): 22–24.

Franklin, Allan. 1990. *Experiment, Right or Wrong*. Cambridge, England: Cambridge University Press.

Franklin, Allan. 1994. "How to avoid the experimenters' regress". *Studies in the History and Philosophy of Science* **25**: 97–121.

Fuller, Steve. 1993. *Philosophy, Rhetoric, and the End of Knowledge: The Coming of Science and Technology Studies*. Madison: University of Wisconsin Press.

Fuller, Steve. 1998. "What does the Sokal hoax say about the prospects for positivism?" To appear in the Proceedings of the International Colloquium on "Positivismes" (Université Libre de Bruxelles and University of Utrecht, 10–12 December 1997), under the aegis of the Académie Internationale d'Histoire des Sciences.

Gabon, Alain. 1994. Review of *Rethinking Technologies*. *SubStance* #**75**: 119–124.

Ghins, Michel. 1992. "Scientific realism and invariance". In *Rationality in Epistemology*, pp. 249–262. Edited by Enrique Villanueva. Atascadero, California: Ridgeview.

Gingras, Yves. 1995. "Un air de radicalisme: Sur quelques tendances récentes en sociologie de la science et de la technologie". *Actes de la recherche en sciences sociales* **108**: 3–17.

Gingras, Yves, and Silvan S. Schweber. 1986. "Constraints on construction". *Social Studies of Science* **16**: 372–383.

Gottfried, Kurt, and Kenneth G. Wilson. 1997. "Science as a cultural construct". *Nature* **386**: 545–547.

Granon-Lafont, Jeanne. 1985. *La Topologie ordinaire de Jacques Lacan*. Paris: Point Hors Ligne.

Granon-Lafont, Jeanne. 1990. *Topologie lacanienne et clinique analytique*. Paris: Point Hors Ligne.

Greenberg, Marvin Jay. 1980. *Euclidean and Non-Euclidean Geometries: Development and History*, 2nd ed. San Francisco: W.H. Freeman.

Gross, Paul R., and Norman Levitt. 1994. *Higher Superstition: The Academic Left and Its Quarrels with Science*. Baltimore: Johns Hopkins University Press.

Gross, Paul R., Norman Levitt, and Martin W. Lewis, eds. 1996. *The Flight from Science and Reason. Annals of the New York Academy of Sciences* **775**.

Grosser, Morton. 1962. *The Discovery of Neptune*. Cambridge, Massachusetts: Harvard University Press.

Guattari, Félix. 1988. "Les énergétiques sémiotiques". In: *Temps et devenir: A partir de l'œuvre d'Ilya Prigogine*, pp. 83–100. Actes du colloque international de 1983 sous la direction de Jean-Pierre Brans, Isabelle Stengers et Philippe Vincke. Geneva: Patiño.

Guattari, Félix. 1995. *Chaosmosis: An Ethico-Aesthetic Paradigm*. Translated by Paul Bains and Julian Pefanis. Bloomington: Indiana University Press. [French original: *Chaosmose*. Paris: Galilée, 1992.]

Harding, Sandra. 1996. "Science is 'good to think with' ". *Social Text* **46/47** (Spring/Summer): 15–26.

Havel, Václav. 1992. "The end of the modern era". *The New York Times* (1 March): E-15.

Hawkins, Harriett. 1995. *Strange Attractors: Literature, Culture and Chaos Theory*. New York: Prentice-Hall/Harvester Wheatsheaf.

Hayles, N. Katherine. 1992. "Gender encoding in fluid mechanics: Masculine channels and feminine flows". *Differences: A Journal of Feminist Cultural Studies* **4**(2): 16–44.

Hegel, Georg Wilhelm Friedrich. 1989 [1812]. *Hegel's Science of Logic*. Translated by A.V. Miller. Foreword by J.N. Findlay. Atlantic Highlands, New Jersey: Humanities Press International.

Henley, Jon. 1997. "Euclidean, Spinozist or existentialist? Er, no. It's simply a load of old tosh". *The Guardian* (1 October): 3.

Hobsbawm, Eric. 1993. "The new threat to history". *New York Review of Books* (16 December): 62–64. [Reprinted in Eric Hobsbawm, *On History*, London: Weidenfeld & Nicolson, 1997, chapter 1.]

Holton, Gerald. 1993. *Science and Anti-Science*. Cambridge, Massachusetts: Harvard University Press.

Hume, David. 1988 [1748]. *An Enquiry Concerning Human Understanding*. Amherst, New York: Prometheus.

Huth, John. 1998. "Latour's relativity". To appear in: *A House Built on Sand: Exposing Postmodernist Myths About Science*, edited by Noretta Koertge. New York: Oxford University Press.

Irigaray, Luce. 1985a. "The 'mechanics' of fluids". In *This Sex Which Is Not One*. Translated by Catherine Porter with Carolyn Burke. Ithaca, N.Y.: Cornell University Press. [French original: *L'Arc*, no. 58 (1974). Reprinted in *Ce sexe qui n'en est pas un*, Paris: Éditions de Minuit, 1977.]

Irigaray, Luce. 1985b. *Parler n'est jamais neutre*. Paris: Éditions de Minuit.

Irigaray, Luce. 1987a. "Le sujet de la science est-il sexué? / Is the subject of science sexed?" Translated by Carol Mastrangelo Bové. *Hypatia* **2**(3): 65–87. [French original: *Les Temps modernes* **9**, no. 436 (November 1982): 960–974. Reprinted in Irigaray 1985b.]

Irigaray, Luce. 1987b. "Sujet de la science, sujet sexué?" In *Sens et place des connaissances dans la société*, pp. 95–121. Paris: Centre National de Recherche Scientifique.

Irigaray, Luce. 1993. "A chance for life: Limits to the concept of the neuter and the universal in science and other disciplines". In *Sexes and Genealogies*, pp. 183–206. Translated by Gillian C. Gill. New York: Columbia University Press. [French original: "Une chance de vivre: Limites au concept de neutre et d'universel dans les sciences et les savoirs". In: *Sexes et parentés*. Paris: Éditions de Minuit, 1987.]

Johnson, George. 1996. "Indian tribes' creationists thwart archeologists". *The New York Times* (22 October): A1, C13.

Kadanoff, Leo P. 1986. "Fractals: Where's the physics?" *Physics Today* **39** (February): 6–7.

Kellert, Stephen H. 1993. *In the Wake of Chaos*. Chicago: University of Chicago Press.

Kimball, Roger. 1990. *Tenured Radicals: How Politics Has Corrupted Higher Education*. New York: Harper & Row.

Kinoshita, Toichiro. 1995. "New value of the α^3 electron anomalous magnetic moment". *Physical Review Letters* **75**: 4728–4731.

Koertge, Noretta, ed. 1998. *A House Built on Sand: Exposing Postmodernist Myths About Science*. New York: Oxford University Press.

Kristeva, Julia. 1969. Σημειωτικὴ: *Recherches pour une sémanalyse*. Paris: Éditions du Seuil.

Kristeva, Julia. 1974. *La Révolution du langage poétique*. Paris: Éditions du Seuil.

Kristeva, Julia. 1977. *Polylogue*. Paris: Éditions du Seuil.

Kristeva, Julia. 1980. *Desire in Language: A Semiotic Approach to Literature and Art*. Edited by Leon S. Roudiez. Translated by Thomas Gora, Alice Jardine, and Leon S. Roudiez. New York: Columbia University Press.

Kuhn, Thomas. 1970. *The Structure of Scientific Revolutions*, 2nd ed. Chicago: University of Chicago Press.

Lacan, Jacques. 1970. "Of structure as an inmixing of an otherness prerequisite to any subject whatever". In *The Languages of Criticism and the Sciences of Man*, pp. 186–200. Edited by Richard Macksey and Eugenio Donato. Baltimore: Johns Hopkins Press.

Lacan, Jacques. 1971. "Position de l'inconscient". In *Écrits 2*, pp. 193–217. Paris: Éditions du Seuil.

Lacan, Jacques. 1973. "L'Étourdit". *Scilicet*, no. 4: 5–52.

Lacan, Jacques. 1975a. *Le Séminaire de Jacques Lacan. Livre XX: Encore, 1972–1973.* Texte établi par Jacques-Alain Miller. Paris: Éditions du Seuil.

Lacan, Jacques. 1975b. Le séminaire de Jacques Lacan (XXII). Texte établi par J.A. Miller. R.S.I. [Réel, Symbolique, Imaginaire] Année 1974–75. Séminaires du 10 et du 17 décembre 1974. *Ornicar?: Bulletin périodique du champ freudien* no. 2: 87–105.

Lacan, Jacques. 1975c. Le séminaire de Jacques Lacan (XXII). Texte établi par J.A. Miller. R.S.I. [Réel, Symbolique, Imaginaire] Année 1974–75. Séminaires du 14 et du 21 janvier 1975. *Ornicar?: Bulletin périodique du champ freudien* no. 3 (May): 95–110.

Lacan, Jacques. 1975d. Le séminaire de Jacques Lacan (XXII). Texte établi par J.A. Miller. R.S.I. [Réel, Symbolique, Imaginaire] Année 1974–75. Séminaires du 11 et du 18 février 1975. *Ornicar?: Bulletin périodique du champ freudien* no. 4 (autumn): 91–106.

Lacan, Jacques. 1975e. Le séminaire de Jacques Lacan (XXII). Texte établi par J.A. Miller. R.S.I. [Réel, Symbolique, Imaginaire] Année 1974–75. Séminaires du 11 et du 18 mars, du 8 et du 15 avril, et du 13 mai 1975. *Ornicar?: Bulletin périodique du champ freudien* no. 5 (winter 1975/76): 17–66.

Lacan, Jacques. 1977a. "Desire and the interpretation of desire in *Hamlet*". Translated by James Hulbert. *Yale French Studies* **55/56**: 11–52.

Lacan, Jacques. 1977b. "The subversion of the subject and the dialectic of desire in the Freudian unconscious". In *Écrits: A Selection*, pp. 292–325. Translated by Alan Sheridan. New York: Norton. [French original: "Subversion du sujet and dialectique du désir dans l'inconscient freudien". In *Écrits*. Paris: Éditions du Seuil, 1966.]

Lacan, Jacques. 1988. *The Seminar of Jacques Lacan. Book II: The Ego in Freud's Theory and in the Technique of Psychoanalysis, 1954–1955.* Edited by Jacques-Alain Miller. Translated by Sylvana Tomaselli with notes by John Forrester. New York: Norton. [French original: *Le séminaire de Jacques Lacan. Livre II: Le Moi dans la théorie de Freud et dans la technique de la psychanalyse, 1954–1955.* Paris: Éditions du Seuil, 1978.]

Lacan, Jacques. 1998. *The Seminar of Jacques Lacan, Book XX, Encore 1972–1973.* Edited by Jacques-Alain Miller. Translated with notes by Bruce Fink. New York: Norton. [French original: see Lacan 1975a.]

Lamont, Michèle. 1987. "How to become a dominant French philosopher: The case of Jacques Derrida". *American Journal of Sociology* **93**: 584–622.

Landsberg, Mitchell [Associated Press]. 1996. "Physicist's spoof on science puts one over on science critics". *International Herald Tribune* (18 May): 1.

Laplace, Pierre Simon. 1995 [5th ed. 1825]. *Philosophical Essay on Probabilities.* Translated by Andrew I. Dale. New York: Springer-Verlag. [French original: *Essai philosophique sur les probabilités.* Paris: Christian Bourgois, 1986.]

Lather, Patti. 1991. *Getting Smart: Feminist Research and Pedagogy With/in the Postmodern.* New York–London: Routledge.

Latour, Bruno. 1987. *Science in Action: How to Follow Scientists and Engineers through Society.* Cambridge, Massachusetts: Harvard University Press.

Latour, Bruno. 1988. "A relativistic account of Einstein's relativity". *Social Studies of Science* **18**: 3–44.

Latour, Bruno. 1995. "Who speaks for science?" *The Sciences* **35**(2) (March–April): 6–7.

Latour, Bruno. 1998. "Ramsès II est-il mort de la tuberculose?". *La Recherche* **307** (March): 84–85. See also errata **308** (April): 85 and **309** (May): 7.

Laudan, Larry. 1981. "The pseudo-science of science?" *Philosophy of the Social Sciences* **11**: 173–198.

Laudan, Larry. 1990a. *Science and Relativism*. Chicago: University of Chicago Press.

Laudan, Larry. 1990b. "Demystifying underdetermination". *Minnesota Studies in the Philosophy of Science* **14**: 267–297.

Lechte, John. 1990. *Julia Kristeva*. London–New York: Routledge.

Lechte, John. 1994. *Fifty Key Contemporary Thinkers: From Structuralism to Postmodernity*. London–New York: Routledge.

Le Monde. 1984a. *Entretiens avec Le Monde. 1. Philosophies*. Introduction de Christian Delacampagne. Paris: Éditions La Découverte et *Le Monde*.

Le Monde. 1984b. *Entretiens avec Le Monde. 3. Idées contemporaines*. Introduction de Christian Descamps. Paris: Éditions La Découverte et *Le Monde*.

Leplin, Jarrett. 1984. *Scientific Realism*. Berkeley: University of California Press.

Leupin, Alexandre. 1991. "Introduction: Voids and knots in knowledge and truth". In: *Lacan and the Human Sciences*, pp. 1–23. Edited by Alexandre Leupin. Lincoln: University of Nebraska Press.

Lévy-Leblond, Jean-Marc. 1997. "La paille des philosophes et la poutre des physiciens". *La Recherche* **299** (June): 9–10.

Lodge, David. 1984. *Small World*. New York: Macmillan.

Lyotard, Jean-François. 1984. *The Postmodern Condition: A Report on Knowledge*. Translated by Geoff Bennington and Brian Massumi. Foreword by Fredric Jameson. Minneapolis: University of Minnesota Press. [French original: *La Condition postmoderne: Rapport sur le savoir*. Paris: Éditions de Minuit, 1979.]

Maddox, John, James Randi and Walter W. Stewart. 1988. " 'High-dilution' experiments a delusion". *Nature* **334**: 287–290.

Maggiori, Robert. 1997. "Fumée sans feu". *Libération* (30 September): 29.

Markley, Robert. 1992. "The irrelevance of reality: Science, ideology and the postmodern universe." *Genre* **25**: 249–276.

Matheson, Carl and Evan Kirchhoff. 1997. "Chaos and literature". *Philosophy and Literature* **21**: 28–45.

Maudlin, Tim. 1994. *Quantum Non-Locality and Relativity: Metaphysical Intimations of Modern Physics*. Aristotelian Society Series, vol. 13. Oxford: Blackwell.

Maudlin, Tim. 1996. "Kuhn édenté: incommensurabilité et choix entre théories." [Original title: "Kuhn defanged: incommensurability and theory-choice."] Translated by

Jean-Pierre Deschepper and Michel Ghins. *Revue philosophique de Louvain* **94**: 428–446.

Maxwell, James Clerk. 1952 [1ˢᵗ ed. 1876]. *Matter and Motion*. New York: Dover.

Mermin, N. David. 1989. *Space and Time in Special Relativity*. Prospect Heights, Illinois: Waveland Press.

Mermin, N. David. 1996a. "What's wrong with this sustaining myth?" *Physics Today* **49**(3) (March): 11–13.

Mermin, N. David. 1996b. "The Golemization of relativity". *Physics Today* **49**(4) (April): 11–13.

Mermin, N. David. 1996c. "Sociologists, scientist continue debate about scientific process". *Physics Today* **49**(7) (July): 11–15, 88.

Mermin, N. David. 1997a. "Sociologists, scientist pick at threads of argument about science". *Physics Today* **50**(1) (January): 92–95.

Mermin, N. David. 1997b. "What's wrong with this reading". *Physics Today* **50**(10) (October): 11–13.

Mermin, N. David. 1998. "The science of science: A physicist reads Barnes, Bloor and Henry". To appear in *Social Studies of Science*.

Miller, Jacques-Alain. 1977–78. "Suture (elements of the logic of the signifier)". *Screen* **18**(4): 24–34.

Milner, Jean-Claude. 1995. *L'œuvre claire: Lacan, la science, la philosophie*. Paris: Seuil.

Moi, Toril. 1986. Introduction to *The Kristeva Reader*. New York: Columbia University Press.

Moore, Patrick. 1996. *The Planet Neptune*, 2ⁿᵈ ed. Chichester: John Wiley & Sons.

Mortley, Raoul. 1991. *French Philosophers in Conversation: Levinas, Schneider, Serres, Irigaray, Le Doeuff, Derrida*. London: Routledge.

Nagel, Ernest and James R. Newman. 1958. *Gödel's Proof*. New York: New York University Press.

Nancy, Jean-Luc and Philippe Lacoue-Labarthe. 1992. *The Title of the Letter: A Reading of Lacan*. Translated by François Raffoul and David Pettigrew. Albany: State University of New York Press. [French original: *Le Titre de la lettre*, 3ᵉᵐᵉ éd. Paris: Galilée, 1990.]

Nanda, Meera. 1997. "The science wars in India". *Dissent* **44**(1) (Winter): 78–83.

Nasio, Juan-David. 1987. *Les Yeux de Laure: Le concept d'objet "a" dans la théorie de J. Lacan. Suivi d'une Introduction à la topologie psychanalytique*. Paris: Aubier.

Nasio, Juan-David. 1992. "Le concept de sujet de l'inconscient". Texte d'une intervention realisée dans le cadre du séminaire de Jacques Lacan "La topologie et le temps", le mardi 15 mai 1979. In: *Cinq leçons sur la théorie de Jacques Lacan*. Paris: Éditions Rivages.

Newton-Smith, W.H. 1981. *The Rationality of Science*. London–New York: Routledge and Kegan Paul.

Norris, Christopher. 1992. *Uncritical Theory: Postmodernism, Intellectuals and the Gulf War*. London: Lawrence and Wishort.

Perrin, Jean. 1990 [1913]. *Atoms*. Translated by D. Ll. Hammick. Woodbridge, Connecticut: Ox Bow Press. [French original: *Les Atomes*. Paris: Presses Universitaires de France, 1970.]

Pinker, Steven. 1995. *The Language Instinct*. London: Penguin.

Plotnitsky, Arkady. 1997. " 'But it is above all not true': Derrida, relativity, and the 'science wars' ". *Postmodern Culture* **7**, no. 2. Available on-line at http://muse.jhu.edu/journals/postmodern_culture/v007/7.2plotnitsky.html

Poincaré, Henri. 1952 [1909]. *Science and Method*. Translated by Francis Maitland. New York: Dover. [French original: *Science et méthode*. Paris: Flammarion, 1909.]

Pollitt, Katha. 1996. "Pomolotov cocktail". *The Nation* (10 June): 9.

Popper, Karl R. 1959. *The Logic of Scientific Discovery*. Translation prepared by the author with the assistance of Julius Freed and Lan Freed. London: Hutchinson.

Popper, Karl. 1974. "Replies to my critics". In *The Philosophy of Karl Popper*, vol. 2, edited by Paul A. Schilpp. LaSalle, Illinois: Open Court Publishing Company.

Prigogine, Ilya, and Isabelle Stengers. 1988. *Entre le temps et l'éternité*. Paris: Fayard.

Putnam, Hilary. 1974. "The 'corroboration' of theories". In *The Philosophy of Karl Popper*, vol. 1, pp. 221–240. Edited by Paul A. Schilpp. LaSalle, Illinois: Open Court Publishing Company.

Putnam, Hilary. 1978. "A critic replies to his philosopher". In *Philosophy As It Is*, edited by Ted Honderich and M. Burnyeat, pp. 377–380. New York: Penguin.

Quine, Willard Van Orman. 1980. "Two Dogmas of Empiricism". In *From a Logical Point of View*, 2nd ed., revised [1st ed. 1953]. Cambridge, Massachusetts: Harvard University Press.

Ragland-Sullivan, Ellie. 1990. "Counting from 0 to 6: Lacan, 'suture', and the imaginary order". In *Criticism and Lacan: Essays and Dialogue on Language, Structure, and the Unconscious*, pp. 31–63. Edited by Patrick Colm Hogan and Lalita Pandit. Athens, Georgia: University of Georgia Press.

Raskin, Marcus G., and Herbert J. Bernstein. 1987. *New Ways of Knowing: The Sciences, Society, and Reconstructive Knowledge*. Totowa, New Jersey: Rowman & Littlefield.

Rees, Martin. 1997. *Before the Beginning: Our Universe and Others*. Reading, Massachusetts: Addison-Wesley.

Revel, Jean-François. 1997. "Les faux prophètes". *Le Point* (11 October): 120–121.

Richelle, Marc. 1998. *Défense des sciences humaines: Vers une désokalisation?* Sprimont (Belgium): Mardaga.

Robbins, Bruce. 1998. "Science-envy: Sokal, science and the police". *Radical Philosophy* **88** (March/April): 2–5.

Rosenberg, John R. 1992. "The clock and the cloud: Chaos and order in *El diablo mundo*". *Revista de Estudios Hispánicos* **26**: 203–225.

Rosenberg, Martin E. 1993. Dynamic and thermodynamic tropes of the subject in Freud and in Deleuze and Guattari. *Postmodern Culture* **4**, no. 1. Available on-line at http://muse.jhu.edu/journals/postmodern_culture/v004/4.1rosenberg.html

Roseveare, N.T. 1982. *Mercury's Perihelion from Le Verrier to Einstein*. Oxford: Clarendon Press.

Ross, Andrew. 1995. "Science backlash on technoskeptics". *The Nation* **261**(10) (2 October): 346–350.

Ross, Andrew. 1996. "Introduction". *Social Text* **46/47** (Spring/Summer): 1–13.

Rötzer, Florian. 1994. *Conversations with French Philosophers*. Translated from the German by Gary E. Aylesworth. Atlantic Highlands, New Jersey: Humanities Press.

Roudinesco, Elisabeth. 1997. *Jacques Lacan*. Translated by Barbara Bray. New York: Columbia University Press. [French original: *Jacques Lacan: Esquisse d'une vie, histoire d'un système de pensée*. Paris: Fayard, 1993.]

Roustang, François. 1990. *The Lacanian Delusion*. Translated by Greg Sims. New York: Oxford University Press. [French original: *Lacan, de l'équivoque à l'impasse*. Paris: Éditions de Minuit, 1986.]

Ruelle, David. 1991. *Chance and Chaos*. Princeton: Princeton University Press.

Ruelle, David. 1994. "Where can one hope to profitably apply the ideas of chaos?" *Physics Today* **47**(7) (July): 24–30.

Russell, Bertrand. 1948. *Human Knowledge: Its Scope and Limits*. London: George Allen and Unwin.

Russell, Bertrand. 1949 [1920]. *The Practice and Theory of Bolshevism*, 2nd ed. London: George Allen and Unwin.

Russell, Bertrand. 1961a. *History of Western Philosophy*, 2nd ed. London: George Allen and Unwin. [Reprinted by Routledge, 1991.]

Russell, Bertrand. 1961b. *The Basic Writings of Bertrand Russell, 1903–1959*. Edited by Robert E. Egner and Lester E. Denonn. New York: Simon and Schuster.

Russell, Bertrand. 1995 [1959]. *My Philosophical Development*. London: Routledge.

Sand, Patrick. 1998. "Left conservatism?" *The Nation* (9 March): 6–7.

Sartori, Leo. 1996. *Understanding Relativity: A Simplified Approach to Einstein's Theories*. Berkeley: University of California Press.

Scott, Janny. 1996. "Postmodern gravity deconstructed, slyly". *The New York Times* (18 May): 1, 22.

Serres, Michel. 1995. "Paris 1800". In *A History of Scientific Thought: Elements of a History of Science*, pp. 422–454. Edited by Michel Serres. Translated from the French. Oxford: Blackwell. [French original: *Eléments d'histoire des sciences*. Sous la direction de Michel Serres. Paris: Bordas, 1989, pp. 337–361.]

Shimony, Abner. 1976. "Comments on two epistemological theses of Thomas Kuhn". In *Essays in Memory of Imre Lakatos*. Edited by R. Cohen *et al.* Dordrecht: D. Reidel Academic Publishers.

Siegel, Harvey. 1987. *Relativism Refuted: A Critique of Contemporary Epistemological Relativism*. Dordrecht: D. Reidel.

Silk, Joseph. 1989. *The Big Bang*, revised and updated ed. New York: W.H. Freeman.

Slezak, Peter. 1994. "A second look at David Bloor's *Knowledge and Social Imagery*". *Philosophy of the Social Sciences* **24**: 336–361.

Sokal, Alan D. 1996a. "Transgressing the boundaries: Toward a transformative hermeneutics of quantum gravity". *Social Text* **46/47** (Spring/Summer): 217–252.

Sokal, Alan. 1996b. "A physicist experiments with cultural studies". *Lingua Franca* **6**(4) (May/June): 62–64.

Sokal, Alan D. 1996c. "Transgressing the boundaries: An afterword". *Dissent* **43**(4) (Fall): 93–99. [A slightly abridged version of this article was published also in *Philosophy and Literature* **20**: 338–346 (1996).]

Sokal, Alan. 1997a. "A plea for reason, evidence and logic". *New Politics* **6**(2) (Winter): 126–129.

Sokal, Alan. 1997b. "Alan Sokal replies [to Stanley Aronowitz]". *Dissent* **44**(1) (Winter): 110–111.

Sokal, Alan. 1998. "What the *Social Text* affair does and does not prove". To appear in *A House Built on Sand: Exposing Postmodernist Myths About Science*, edited by Noretta Koertge. New York: Oxford University Press.

Stengers, Isabelle. 1997. "Un impossible débat". Interview with Eric de Bellefroid. *La Libre Belgique* (1 October): 21.

Stove, D.C. 1982. *Popper and After: Four Modern Irrationalists*. Oxford: Pergamon Press.

Sussmann, Hector J., and Raphael S. Zahler. 1978. "Catastrophe theory as applied to the social and biological sciences: A critique". *Synthese* **37**: 117–216.

Taylor, Edwin F. and John Archibald Wheeler. 1966. *Spacetime Physics*. San Francisco: W. H. Freeman.

University of Warwick. 1997. "DeleuzeGuattari and Matter: A conference". Philosophy Department, University of Warwick (UK), 18–19 October 1997. Conference description available on-line at http://www.csv.warwick.ac.uk/fac/soc/Philosophy/matter.html

Van Dyck, Robert S., Jr., Paul B. Schwinberg, and Hans G. Dehmelt. 1987. "New high-precision comparison of electron and positron *g* factors". *Physical Review Letters* **59**: 26–29.

Van Peer, Willie. 1998. "Sense and nonsense of chaos theory in literary studies". In *The Third Culture: Literature and Science*, pp. 40–48. Edited by Elinor S. Shaffer. Berlin–New York: Walter de Gruyter.

Vappereau, Jean Michel. 1985. *Essaim: Le Groupe fondamental du nœud*. Psychanalyse et Topologie du Sujet. Paris: Point Hors Ligne.

Vappereau, Jean Michel. 1995. "Surmoi". *Encyclopaedia Universalis* **21**: 885–889.

Virilio, Paul. 1984. *L'Espace critique*. Paris: Christian Bourgois.

Virilio, Paul. 1989. "Trans-Appearance". Translated by Diana Stoll. *Artforum* **27**, no. 10 (1 June): 129–130.

Virilio, Paul. 1990. *L'Inertie polaire*. Paris: Christian Bourgois.

Virilio, Paul. 1991. *The Lost Dimension*. Translated by Daniel Moshenberg. New York: Semiotext(e). [French original: see Virilio 1984.]

Virilio, Paul. 1993. "The third interval: A critical transition". Translated by Tom Conley. In *Rethinking Technologies*, pp. 3–12, edited by Verena Andermatt Conley on behalf of the Miami Theory Collective. Minneapolis: University of Minnesota Press.

Virilio, Paul. 1995. *La Vitesse de libération*. Paris: Galilée.

Virilio, Paul. 1997. *Open Sky*. Translated by Julie Rose. London: Verso. [French original: see Virilio 1995.]

Weill, Nicolas. "La mystification pédagogique du professeur Sokal". *Le Monde* (20 December): 1, 16.

Weinberg, Steven. 1977. *The First Three Minutes: A Modern View of the Origin of the Universe*. New York: Basic Books.

Weinberg, Steven. 1992. *Dreams of a Final Theory*. New York: Pantheon.

Weinberg, Steven. 1995. "Reductionism redux". *New York Review of Books* **42**(15) (5 October): 39–42.

Weinberg, Steven. 1996a. "Sokal's hoax". *New York Review of Books* **43**(13) (8 August): 11–15.

Weinberg, Steven *et al.* 1996b. "Sokal's hoax: An exchange". *New York Review of Books* **43**(15) (3 October): 54–56.

Willis, Ellen. 1996. "My Sokaled life". *Village Voice* (25 June): 20–21.

Willis, Ellen *et al.* 1998. "Epistemology and vinegar". [Letters in response to Sand 1998.] *The Nation* (11 May): 2, 59–60.

Zahler, Raphael S., and Hector J. Sussmann. 1977. "Claims and accomplishments of applied catastrophe theory". *Nature* **269**: 759–763.

Zarlengo, Kristina. 1998. "J'accuse!" *Lingua Franca* **8**(3) (April): 10–11.

Index